ELECTRIC LIGHT

ELECTRIC LIGHT

AN ARCHITECTURAL HISTORY

Sandy Isenstadt

THE MIT PRESS
CAMBRIDGE, MASSACHUSETTS
LONDON, ENGLAND

© 2018 Massachusetts Institute of Technology

All rights reserved. No part of this book may be reproduced in any form by any electronic or mechanical means (including photocopying, recording, or information storage and retrieval) without permission in writing from the publisher.

This book was set in PFDIN Pro by The MIT Press. Printed and bound in the United States of America.

Library of Congress Cataloging-in-Publication Data
Names: Isenstadt, Sandy, 1957- author.
Title: Electric light : an architectural history / Sandy Isenstadt.
Description: Cambridge, MA : The MIT Press, [2018] | Includes bibliographical references and index.
Identifiers: LCCN 2017053679 | ISBN 9780262038171 (hardcover : alk. paper)
Subjects: LCSH: Electric lighting--Psychological aspects. | Electric lighting--Social aspects. | Municipal lighting--History. | Street lighting--History. | Lighting, Architectural and decorative--History.
Classification: LCC TK4188 .I84 2018 | DDC 621.3209--dc23 LC record available at https://lccn.loc.gov/2017053679

10 9 8 7 6 5 4 3 2 1

To Hannah   fuel spark flame   circle of light

CONTENTS

Acknowledgments ix

1   THE ARCHITECTURE OF ELECTRIC LIGHT   1
2   AT THE FLIP OF A SWITCH   24
3   DRIVING THROUGH THE AMERICAN NIGHT   66
4   LIGHTING FOR LABOR   106
5   ELECTRIC SPEECH IN THE CITY   154
6   GROPING IN THE DARK   206

Notes   243
Index   279

Acknowledgments

I have benefited immeasurably from the advice, insight, and encouragement of brilliant and generous colleagues. For taking the time to talk with me about various ideas for this book and proposing stimulating questions, I would like to thank Jenny Anger, Ann Ardis, Wendy Bellion, Ruth Bielfeldt, Yve-Alain Bois, Martin Brückner, Siobhan Carroll, Monica Dominguez Torres, Alice Friedman, Richie Garrison, Jana Hatáková, Jason Hill, Paul Jackson, Nina Athanassoglou-Kallmyer, Juliet Koss, Susan Laxton, Michael Lewis, Larry Nees, Wallis Miller, Kristen Poole, Stefan Schmitt, Jeffrey Schnapp, Mitchell Schwarzer, Oliver Scheiding, David Suisman, Larry Vale, David van Zanten, Heinrich von Staden, Gennifer Weisenfeld, Margaret Werth, Mary Woods, Gwendolyn Wright, and Bernie Zirnheld.

I am grateful to Jeffrey Cohen, Aaron Wunsch, and Carla Yanni for closely reading draft chapters and rendering smart criticism, and I am especially appreciative of my colleagues in electric light: Holly Clayson, Tim Edensor, Dietrich Neumann, David Nye, Margaret Maile Petty, Wolfgang Schivelbusch, and Hélène Valance. My thanks to all of you for your comments and conversations over the years. In particular, I thank Ken Kalfus, Natalie Shivers, and Sarah Wasserman for their time, effort, and thoughtful editorial suggestions. This book would be all the lesser without your help. I am happily indebted to Hannah Bennett for her heroic efforts gathering images and permissions from the darkest corners of the world of lighting ephemera. I am likewise thankful to the MIT Press—Roger Conover, Deborah Cantor-Adams, Margarita Encomienda, Gabriela Bueno Gibbs, and Marjorie Pannell, in particular—for their editorial and design guidance.

I am honored to have received support for this project from several programs and institutions at various stages of its development. I offer my sincere thanks in return to the Edison Price Fellowship of The Nuckolls Fund for Lighting Education, the Griswold Research Fund at Yale University, the Institute for Advanced Study, Princeton, New Jersey, the Center for Material Culture Studies, the Art History department, and the Office of the Dean at the University of Delaware.

1
THE ARCHITECTURE OF ELECTRIC LIGHT

> Space no longer exists: the street pavement,
> soaked by rain beneath the glare of electric lamps,
> becomes immensely deep and gapes to the
> very centre of the earth.
> **Umberto Boccioni and his Futurist colleagues, 1910**[1]

Electric lighting brightens the world. While earlier forms of artificial lighting helped people maneuver, work, and play after sunset, electricity brought with it an utterly new era of nighttime activity. At the flick of a switch, electric light blanches the night, banishing nocturnal gloom to the extent that one can easily forget that light was once a miracle of biblical proportion. Today, in countless cities across the globe, electric light is readily available, usually cheap, and generally reliable. Lighting the built environment is commonplace, seemingly an inevitable step forward in humankind's technological progress. The alternation of day and night, the transcendent frame for life on Earth, is now provisional; light, lots of it, is swiftly summoned to serve production, convenience, and amusement. The entire planet has by now been mantled in colors and a luminous character hardly envisaged little more than a hundred years ago. Naturalized in its ubiquity, its benefits countless, its bright, crisp, visual character the very image of optimism and reason, electric light has become inextricable from modern life and, at the same time, unremarkable.

Electric light is presumed simply to illuminate volumes and masses that exist whether or not the lights are on. It brightens a room or street but is not imagined to change the underlying nature of a space. This book argues that it does. The change in the perception of spaces brought about by different forms of electric light was tantamount to an actual change in the character, use, and certainly the understanding of space. This claim is not based in phenomenology or a description of how nocturnal volumes are limned into perceptual presence by light. Rather, it is based on operative or behavioral responses to the changed visual character wrought by electric light, on how people apprehended and acted within spaces as those spaces were perceptually transformed. The primary claim here is that the introduction of electric lighting in the late nineteenth and early twentieth centuries led to the formation of new luminous spaces and a new set of visual conditions for ordinary people to negotiate. Far more than simply ornamental or a means to brighten preexisting rooms and surfaces, electric light generated unprecedented spaces that altered and often eclipsed other, prior spaces.

Electric light, in short, is a form of architecture; it is a new kind of building material, a set of evolving compositional practices, and a host of occupational strategies that produced new places to live, work, and play. It is an instrument of spatial invention as well as a decisive condition for nighttime inhabitation. This book aims to develop an architectural history based on the shaping and cultural assimilation of new perceptual conditions. To describe these effects, this book considers several cases to suggest an architectural history of spaces that have been generated or extensively reconstituted by electric light.

Although electrification was a momentous development that transformed nearly every facet of American society, it has sparked little interest among historians of architecture until recently. While the field has changed considerably in the last thirty or so years, it was long focused on variations in the form of buildings and landscapes as a result of developments in construction technologies, materials, social relations, economic pressures, design thinking, and so on. Modernism in architecture, for example, had long been understood as an attempt to construct buildings that were formally proportionate to the industrial origins of their materials and to the technologically advanced methods of their assembly. Such a mandate of technological commensurability was for advocates of modernism a lever to lift modern work above the fray of stylistic battles that was believed to have exhausted architecture in the twentieth century. Accordingly, histories of modern architecture have often focused on formal inventions such as exposed steel and window walls, in which the products of manufacturing processes, such as rolled steel or float-processed glass, are given visual prominence. Historians have tended to concentrate on those architects who devised new compositional means to exploit and aesthetically emphasize such materials or used them to further new ideas about architectural form. Although the development and subsequent spread of electric light were clearly important technological advances, architects were slow to alter their practice in response, and only lately have historians considered the effect of easily accessible electric lighting on architecture. This book is indebted to recent scholarship that has opened up architectural history to encompass hitherto overlooked aspects of modernity, and more specifically to those authors who have focused on questions regarding the cultural transformations consequent on electrification.

ELECTRIFICATION

The extraordinarily rapid development of electric lighting does not seem to have been brought home to the mind of the general public.
Alfred Hay, 1897[2]

Demand for better sources of artificial light increased with the economic expansion of the European economy in the eighteenth century. Patents and competitions for improved lamps, such as the 1780 Argand oil lamp, had risen steeply by the century's end.[3] New, more refined oils and distillates came to market, although many were unstable or had low flash points that made them unsuited for mills with airborne lint. Finer fuels were regulated and taxed, but even sales of poor-quality tallow candles increased fivefold from 1750 to 1800.[4] Subsequent lighting developments included new fuels such as stearine and paraffin, plaited and chemically treated wicks, and new fixtures with greater adjustability, more highly polished reflecting surfaces, and pumps for feeding fuel to the wick.[5] Finally, John Walker, an English chemist, developed an easy-to-use friction match around 1826 by coating fine wooden sticks with sulfur, the noxious smell of which likely led to the common name of "Lucifers." Walker's friction match was later improved by others with the addition of a phosphorus coating (figure 1.1).[6]

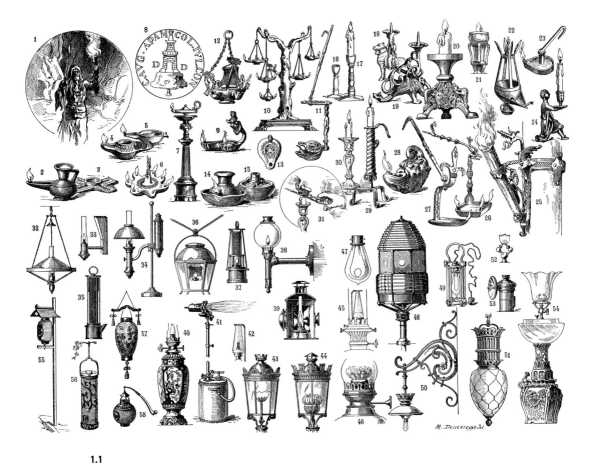

1.1
Progress in lighting from antiquity to 1900.
"Eclairage," *Nouveau Larousse Illustré*, vol. 4, ed. Claude Augé (Paris: Libraire Larousse, 1900), 32.

1.2
Yablochkov candles illuminate the Avenue de l'Opera. "L'Avenue de l'Opéra éclairée par les lampes Jablochkoff," *Lumière Électrique: Revue Universelle d'Électricité* 4, no. 38 (Aug. 10, 1881): 186.

The history of electric lighting has been told often enough to need only the briefest review here. It is generally traced to the British chemist Humphry Davy and his early nineteenth-century electrolysis experiments employing incandescent and arc lighting, illumination resulting from the heating of a filament and illumination resulting from creating an arc of electricity between two carbon rods or electrodes, respectively. While many researched the potential for electric lighting in the following decades, the first commercially practicable arc lamp was invented only in 1876, by the Russian electrical engineer Pavel Yablochkov, who then traveled to Paris and installed the lamps in a department store in 1877 and along the Avenue de l'Opéra and the Place de l'Opéra in 1878 for the Third Exposition Universelle (figure 1.2). Within a short time, many cities in Europe had mounted Yablochkov candles, as they came to be known, on rows of poles along their major streets. In the American West a number of towns, such as San Jose, Austin, and Los Angeles, clustered similar arc lamps atop tall masts to light their main streets at night.[7] Arc lamps were very bright and best suited for outdoor uses such as street lighting or for lighting large interiors such as factories or department stores. Regardless, in 1878 William Armstrong, an English scientist and industrialist, installed one in his home, Cragside, in Northumberland, most likely the first use of electric light in a domestic setting. It was powered by a small hydroelectric plant on the estate and replaced two years later by incandescent bulbs manufactured by Joseph Swan.

By that time Thomas Edison had patented his own incandescent bulb, and in 1882 he built the first large-scale power station, at Pearl Street in downtown New York City, establishing principles for the commercial distribution of direct current electricity from a central power station. Prior to this, municipalities or building owners had to generate their own electricity. In 1886 George Westinghouse, the American engineer and electrical entrepreneur, set up the first alternating current distribution system, which was far better suited than direct current for long-distance transmission and, as a result, eventually drove Edison's direct current stations out of business. As other central power stations adopted alternating current technology, they exploited economies of scale by interconnecting with one another to share capacity. This was the start of a national electricity grid, a creation that in 2003 the National Academy of Engineering named the greatest engineering achievement of the twentieth century. With innovators such as the entrepreneur Samuel Insull, who developed metered demand pricing rather than flat-rate billing and marketed electricity well beyond its primary use for lighting, and with the manufacturing of cheaper, brighter, and more durable bulbs, as well as the implementation of government programs such as the 1933 Tennessee Valley Authority to provide electricity to underserved regions, access to electricity spread throughout

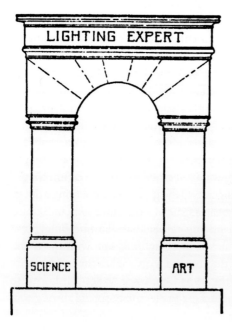

1.3
"Electric Light's Golden Jubilee" commemorative two-cent stamp, issued in 1929 by the U.S. Post Office, to celebrate the fiftieth anniversary of Thomas Edison's demonstration of the incandescent light bulb.

1.4
Art and science, twin pillars of the lighting expert. Matthew Luckiesh, "Linking Science and Art with Practice in Lighting," *TIES (Transactions of the Illuminating Engineering Society)* 12 (February 1917): 97–98.

the nation. In 1879 electric light was a luxury, and a new row of lamps was a newsworthy event. In 1929, at the time of the electrical industry's celebration of "Electric Light's Golden Jubilee"—fifty years after Edison's demonstration of his incandescent lamp—even modest homes could be expected to have at least several bulbs burning at night (figure 1.3). Even in the dark days of the Great Depression, electric light remained an undimmed promise of material progress in America.[8]

The spread of electric lighting, along with the new competitiveness of gas lighting owing to the invention of the incandescent Welsbach gas mantle, the low price of kerosene as an illuminant, and, starting in the 1890s, the use of acetylene lamps, spurred the formation of professional lighting societies. An International Photometric Commission was established in 1900 at the Paris Exposition, then enlarged in 1913 and renamed the International Commission on Illumination. The American Illuminating Engineering Society (IES) was founded in 1906 and the British Illuminating Engineering Society was started three years later. Until that time, various parties, such as central stations, manufacturers of electric lighting equipment, electrical engineers, a cadre of variously trained lighting consultants, and architects, had had the most say over the design of electric lighting systems. The IES founders argued that the field had evolved rapidly in recent years and a new profession dedicated to lighting was warranted.In keeping with Progressive Era concerns for worker safety, a primary concern of the IES had to do with efficiency and waste, which, as the historian Margaret Maile Petty points out, entails human as well as economic costs. At stake was the nation's eyesight, as well as tens of millions of dollars lost to poorly designed lighting.[9] At the same time, the IES aimed to address a broader public on matters of illumination. In expounding their professional rationale, the IES founders argued that illumination and social progress moved forward in tandem. Done well, lighting fostered material progress for all those who labored or leisured by artificial light.[10]

The formation of the IES initially caused tension with the vastly larger and more established American Institute of Electrical Engineers, which sought to keep the newly named "illuminating engineer" as a subset within the engineering profession. Some IES members even argued that the term "engineer" should not appear in the society's name since it suggested a merely procedural orientation.[11] "Lighting expert," they said, would better represent their concern with human physiology and psychology, architecture and interior design principles, and manufacturing methods, as well as the physics of artificial light sources. The hybrid mission to address both "the Science and Art of illumination" was written into the IES constitution (figure 1.4).[12]

With that, the IES also found itself in tension with architects, who were often an important clientele. Although seeking common ground with architects,

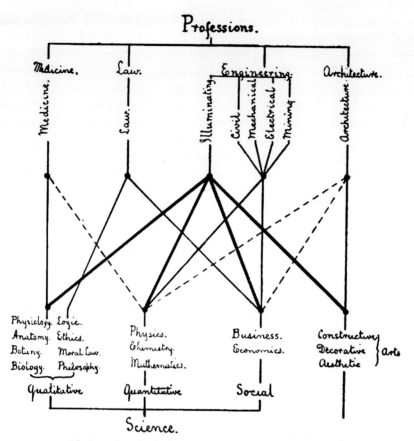

1.5
Diagram depicting illuminating engineering's unique position among the professions, requiring breadth of knowledge in the natural sciences, the social sciences, and the arts. Arthur Edwin Kennelly, "The Profession of Illuminating Engineering: A Presidential Inaugural Address," *TIES (Transactions of the Illuminating Engineering Society)* 6, no. 1 (January 1911): 76.

E. Leavenworth Elliott, a cofounder of the IES and editor of *The Illuminating Engineer*, revealed his impatience at points and repudiated architects' claim on aesthetic questions. In 1907, following a century of industrialization in building construction, Elliott wrote that architects simply dressed up engineered structures with stale historical motifs: "The architect of the present time is a mere copyist and compiler."[13] Or, as he put it sharply a few months later, architects were "the *bete noire* of illuminating engineering,—and it must be admitted, with a considerable degree of reason."[14] Both professions agreed that electric light altered visual conditions, but their disciplinary perspectives were mismatched; the engineer's ideas of utility were often at odds with the architect's ideas of beauty (figure 1.5).

For their part, many American architects practicing at the turn of the twentieth century were alarmed by the threat electric light posed to the proprieties and conventions of their profession. Poorly placed lighting destroyed the delicate poise architects tried to achieve with their designs; the new spaces limned by light could visually disrupt otherwise carefully composed volumes. Architects were more likely to worry about a lighting fixture's stylistic fit within a room than whether or not it provided light adequate to a specific purpose. Tradition-minded architects might even recommend that their classically inspired buildings not be lighted at night since doing so would obliterate the balance of forms that characterized that style. Some astute lighting designers advocated that colleagues learn the formal principles of architecture to better understand such concerns and develop mutually satisfactory schemes.[15] In time, and after a great deal of debate, the two professions learned to work together to produce appealing and intelligently lighted buildings.[16] In addition to working as consultants on building projects, the IES commissioned research from its members, published a journal, developed lighting standards for a wide range of occupations, and lobbied legislatures to introduce state and federal lighting standards.

AN ELECTRIC MODERNISM

Huge wars, great inventions, conquest of the air, speed of travel,
telephones, telegraphs, dreadnoughts—the realm of electricity.
But our young artists paint Neros and half-naked Roman warriors.
Kazimir Malevich, 1916[17]

The creation and spread of fixed infrastructures of artificial light—gas first, then electric—was as fundamental to the making of the modern world as any system of transportation, communication, or energy and as momentous as industrial urbanization itself. But for some, electric light was more than just another modern invention: it pointed toward an emerging way of understanding, inhabiting, and

engaging aesthetically with a technologically transformed world. It entailed in itself a kind of modernism. By making the benefits of technological achievement visually and vividly immediate, electric lighting reinforced—indeed, seemed to deliver on—the progressive promises of modernity. For this reason, it appeared as an iconic element in artistic representations of modern life in the early twentieth century. It conveyed the dawning ecstasy of a new social sublime, such as the gleaming glass architecture of the German visionary author Paul Scheerbart, the radiant Alpine cities of avant-garde German architect Bruno Taut, or the scorching street lamp of the Italian Futurist Giacomo Balla (figure 1.6).

In this larger context, electric light, with its motility and lability, was as restless and homeless as any character from James Joyce, as international as a restive émigré, as disinterested as Ulrich in Austrian writer Robert Musil's unfinished novel, *The Man without Qualities*, as visually splintered as a Cubist canvas, and often as optically dissonant as Arnold Schoenberg was atonal. In many ways, electric light manifested in the lived world what other arts tried to evoke within their specific media: a new kind of vision that was calibrated to the technology and turbulence of modern society. Modernism's revolutionary poet, Vladimir Mayakovsky, extolled New York City's "constant electrical breeze" when he visited there in 1925: "The buildings are glowing with electricity,"[18] he wrote. Ablaze with electric lamps, the city's streets seemed to him to be the native habitat of the fevered mind of modernity, a place where even avant-garde attitudes of the Old World boiled away in the crackling white-hot atmosphere of Broadway. As a poet seeking new forms of life and finding them in lower Manhattan, Mayakovsky declared, "I am in debt to Broadway's dazzling streetlamps."[19] If modernity itself can be characterized by rapid, incessant change—and modernism as the creative and conscious response to such change—then electric light—instantaneous, malleable, ubiquitous, evanescent—is modernity's medium.

1.6

Giacomo Balla, *Street Light*, ca. 1910–1911 (dated on painting 1909). Balla, a Futurist painter committed to the embrace of modernity, depicted the electric lamp's eclipse of natural sources of light, such as the Moon, and emphasized the lamp's glare as central to its fascination. © 2017 Artists Rights Society (ARS), New York/SIAE, Rome.

While familiar markers of modernism in architecture have included a rhetoric of historical rupture, a functionalist credo, new relations of labor and capital, media saturation, and tectonic representation and abstraction, the term is broadened here to include new formations of urban experiences, such as those Mayakovsky rendered. Modernism here involves the conscious recognition—sometimes in celebration, sometimes in alarm—of breaches triggered by new technologies in the custom and scale of everyday life. It embraces, in particular, changes to the visual environment that were largely innocent of an explicit aesthetic agenda but that resulted nonetheless in some of the most spectacular new spaces in the American landscape. Utilitarian rationales, both convincing and banal, ushered into everyday life substantial reconfigurations of sensory experience.

Electric Light: An Architectural History thus attempts to articulate an "electric modernism" grounded in the formation of new sorts of spaces unique to an era of electric light. It conceives modernism in architecture to include not just the aesthetic exploitation of new materials and methods of construction but also new perceptual conditions and new visual habits. Electric lighting came to illuminate structures that were built well before its invention, but it also produced new spaces that were to a large extent defined by and effectively comprised of light. It motivated the close study of vision by a profession of lighting engineers. It empowered ordinary people with the ability to alter their visual environment in a flash. It sparked new optical metrics for work and for leisure, for production and for consumption. It generated new visual categories for daily routines, led to a new geography of night, and fueled new expectations regarding the character, amenities, and potential of the nighttime environment.

A history of electric modernism shares a number of traits with more familiar sorts of histories of modernism, including an acceptance of the shaping influence of technological change on aesthetic expression in the industrial age, a recognition of commercial and ideological motivations underpinning the formation of the built environment, and concurrence with a brand of epiphenomenalism that finds aesthetic effects issuing from strictly utilitarian motivations regarding, say, function and construction. But it also very nearly doubles what might be said about the nature of architectural and urban form: architectural criticism and much scholarly study have implicitly presumed a building washed in daylight, as if architecture's fullest aesthetic and even historical consideration can occur only while the sun is up. Indeed, the spaces generated by electric light—the swinging cone projected from a handheld flashlight, for example—do not even register as spatial inventions in secondary work, even though they appear in contemporaneous accounts. With daylight generally posited to be the normal way to see and understand architecture and the city, half the diurnal cycle—the night—is ignored. This book adds to the recent literature attempting to address this oversight.

1.7

Maxfield Parrish, *Prometheus*, oil on panel, 1919. Edison was often compared to Prometheus, who stole light from the gods for the use of humanity. The image recapitulates the mystic origin myth of electric light that was common in the early decades of electric lighting. The work was created for General Electric calendars advertising Edison Mazda Lamps, 1919.

LIGHT WITHOUT FLAME

Behind the screens the light was shining, nice plain electric light without any wicks to trim or chimneys to break. That was a relief anyhow. ... To the older peoples, electric light, ice factories and shower baths came at the end of things, not at the beginning.
Winifred Lewellin James, 1914[20]

While some of the effects of electric light were evident with earlier forms of illumination, it was strikingly distinct in many ways. In an overview aimed at a general audience, Franklin Pope, an electrical engineer, telegrapher, and one-time collaborator of Thomas Edison, wrote that the new light bulb emitted "a light that produces no deleterious gases, no smoke, no offensive odors—a light without flame, without danger, requiring no matches to ignite, giving out but little heat, vitiating no air, and free from all flickering; a light that is a little globe of sunshine, a veritable Aladdin's lamp" (figure 1.7)[21] These traits were all routinely measured against those of gas, which produced soot and sometimes odors, raised room temperatures, and flickered from impurities in fuel or wick or with the movement of air. As numerous people noted at the time, electricity provided light without flame, an unprecedented separation of phenomena that had been paired in ordinary experience since prehistory. Electric light was clean and safe, with no valves to twist, no tinder boxes or matches to strike, and no undetected buildup of explosive fuel. Moreover, it competed in cost with other lighting technologies such as gas or kerosene lamps.

As a technological advance, electric light was something even more. It exhibited two remarkable and related aspects: instantaneity and action at a distance (the idea that one object can affect another without physical contact), features of electricity more generally. How objects affect one another has been an abiding concern of myth and science since the dawn of history, with the usual presumption being that two things must somehow touch in order to transfer force. Early atomistic and mechanistic theories of matter were premised almost exclusively on contacts and collisions between solids. Cause and effect were bound by contiguity. Newton struggled with this conundrum when he contemplated the question of gravity, as had Lucretius before him; Lucretius held that invisible forces were exerted across great distances. Investigations into the nature of electricity and magnetism in the late eighteenth and early nineteenth centuries likewise compelled a theory to explain these unseen forces, which often led to a vague notion of action at a distance. The British physicist Michael Faraday, followed by the Scottish physicist James Maxwell, in their investigations of electromagnetic induction proposed a field of physical forces, an intervening medium that mediated the transfer of energy. As Maxwell concluded in his treatise on the subject, "Whenever energy is transmitted from one body to another in time, there must be a medium or substance in which the energy exists after it leaves one body and before it reaches the other." Thus, he continued, space is not empty but filled with forces, and all theories regarding action at a distance "lead to the conception of a medium in which the propagation [of energy] takes place."[22] The embrace of Faraday and Maxwell's field theory and its subsequent modifications mooted the question of action at a distance for scientists until the emergence of quantum physics.

In the 1840s, telegraphy, powered by electricity, seemed to make instantaneity and action at a distance everyday occurrences. With a single telegraphic transmission between one point and another, space itself seemed to be compressed; mountains, plains, deserts, rivers, oceans—distance and its topographic inflections—became irrelevant. As space seemed to collapse, so too did duration diminish, seemingly down to zero: people learned about distant events at the very moment events were unfolding since transmissions could go out to infinite points from a single origin. While scientists were able to explain these phenomena matter-of-factly, such demonstrations of compressed space and collapsed time seemed a wonder, especially in relation to mechanical ideas of cause and effect that were rational and practical presumptions for the conduct of daily life.[23] As much as it was a handy communications tool, telegraphy was also a striking demonstration of humankind's mastery over time and space.

The immediate illumination of the electric light, achieved with the flick of a little switch, is clearly related to the simultaneity of the telegraph in its instantaneity and its seeming ability to act at almost any distance. But even more than the telegraph, the light switch, commonplace not long after its introduction, made the experience of instantaneity and action at a distance both more intimate and more vivid. Whereas the telegraph abrogated physical geography in its creation of a universal time—the now—the light switch instantiated variable geographies each of whose contours would be determined by the scope of just released light. With the telegraph, the present instant, which included distant news absorbed within the immediate environment, became a new historical object; with the electric light switch, that present instant acquired a historically unique and powerfully visible form. What is seen when the button is pressed is not just so many newly illuminated things but a spectacle of instantaneity and a perceptible token of an age of acceleration. If the telegraph effectively deterritorialized information to create a global regime of information, the light switch variably reterritorialized vision to create countless individually willed districts of vision. It also made instantaneity and action at a distance, formerly chimera of theoretical physics, features of daily life.

1.8
Glare is as much a matter of placement as absolute brightness. Here, "serious glare" is created by a lamp placed at the curve of a country road, obscuring cars approaching from beyond the curve. Preston Millar, "The Effective Illumination of Streets," *Transactions of the American Institute of Electrical Engineers* 35 (June 1915), plate 78.

1.9
"To spend an evening reading in the general reading room of the Congressional Library is, optically speaking, to take one's life in his hands." Photograph from E. Leavenworth Elliott, "The Congressional Library: The American Chamber of Horrors in Illumination," *Illuminating Engineer* 5, no. 10 (December 1910): 499.

THE EVIL OF GLARE

A new sort of urban star now shines out nightly, horrible, unearthly,
obnoxious to the human eye; a lamp for a nightmare! Such a
light as this should shine only on murders and public crime, or along
the corridors of lunatic asylums, a horror to heighten horror.
Robert Louis Stevenson, 1878[24]

Electric light was also unique in bringing with it what was arguably the greatest visual problem of the late nineteenth and early twentieth centuries: glare. To the considerable extent that vision informs spatial experience, glare also became a significant problem for architecture, in the broadened sense followed here. Whether brought on by the flash of oncoming headlights or a light being switched on, glare disrupted not only the ability to see clearly (figure 1.8). Glare disrupted, however briefly, the mental representation of space and, despite the modern age's accelerated sense of time, measurably deferred spatial recognition. It introduced into a modernist drive for immediacy the physiological time of ocular adaptation, a time period ranging from seconds to minutes for the eyes to adjust to sharp differences in lighting levels. Under most conditions, this duration was negligible, barely noticed, as memory or gradually sharpening outlines sufficed to guide movement until vision was fully regained. At other times, however, the sudden blindness of glare could mean the difference between safety and injury, even between life and death, as spatial perception was distorted long enough to facilitate a misstep or misplaced hand or a mistaken turn. At still other moments, when flashing or bright lights were the object of perception, such as at expositions, glare became a leading element of spatial experience. Its evanescence and inconstancy and the delightful havoc it played with vision made the borders of a place equivocal and rendered it a trembling volume. In recognition of its ubiquity and the challenges it posed to illuminating engineers, the glare problem runs throughout the subsequent chapters of this book as a source of both visual noise and dire peril.

"Glare" has a range of meanings, from the marked difference between the luminous intensity of an object of vision and an intervening source that makes seeing difficult, to an unpleasant sensation resulting from excessive brightness, or what today would be called disability glare and discomfort glare, respectively. Numerous factors bear on the experience of glare. Some can be readily measured, such as the brightness of the light source relative to the light reflected off the object of vision, the relative positions of the source and object in the field of view, the total volume of light entering the eye, the length of exposure to the luminous field, the degree of contrast with a background, and so on. Such variables help explain why, for example, automobile headlights that produce glare by night might go unnoticed during the day. But glare also depends on the viewer, the condition of the retina, the viewer's physiological profile at a particular moment, whether the viewer is trying to read something close by or discern a distant object—in other words, virtually anything that can affect whether or not something will be seen well (figure 1.9). As such, glare is to a large degree subjective, something plainly evident in the way lighting engineers have historically described it: "Any brightness within the field of vision of such a character as to cause discomfort, annoyance, interference with vision, or eye fatigue."[25] As Charles Steinmetz, one of the great pioneers of alternating current theory, put it regarding light more generally: "Light is not a physical quantity, but is the physiological effect exerted on the human eye by certain radiations within a certain narrow range of frequency."[26] Accordingly, with body and environment both contributing factors, glare for many years was investigated along two tracks, according to the reported visibility of a test object under varying lighting conditions and according to viewers' own estimates of the "unpleasantness, discomfort, shock, dazzle, etc." caused by a glaring light.[27] To this day, and despite remarkable advances in lighting design, technology, and measurement, negotiating glare remains a challenge.

Glare was hardly a modern phenomenon: sunlight reflected off water, for example, could be glaring if it interfered with an effort to see something in particular. Conversely, even a relatively dim light source, such as a candle, could be glaring if it was situated within the field of vision of someone trying to see a more dimly lit object behind or beside it. But electric lighting made glare a national, indeed, an international, concern at the start of the twentieth century. Electric lighting was many times brighter than lighting from prior sources and the resulting intensity of illumination posed a problem, especially for those unaccustomed to such brilliance.[28] In addition, the urban environment rapidly began to contain more and brighter sources of artificial light as electric lighting, initially costly, declined steeply in price in the decades following the turn of the twentieth century. With advisory groups such as the IES just forming, there were few established protocols relevant to the numerous conditions in which electric light would soon be installed. Lamps were often mounted by people with little practical experience in doing so. The lamps themselves were initially adapted from the designs of gas lamps, which shone a fraction as brightly as electric lamps. As a result, the nighttime visual field of the early twentieth century was punctuated by uncomfortable episodes of optical strain. It was encountered on city streets as well as country lanes; it was both nuisance and hazard in the workplace and impeded economic growth. At school and at home, glare disrupted productive attention and troubled quiet and leisure.

To some extent, glare was part of the initial fascination with electric light. It was a novel sensation. In his 1883 story, *The Ladies' Paradise,* Zola likened electric lights to the new manufactured goods available to the coalescing class of middle-income female shoppers: the lamps' "white brightness of a blinding fixity" replaced twilight with a "colossal orgie of white [that] was also burning, itself becoming a light."[29] In their novelty and vividness, electric lights and consumer goods were juggernauts of the modern age. Theaters in the 1880s, even those still lit by gas, experienced an "incipient light cult" evident in higher lighting levels established by electric lighting.[30] As electric lighting spread to the domestic sphere, homeowners might mistake brightness for being up-to-date and have overly bright bulbs installed. There was nothing inherent in electric lighting to cause glare, experts noted.[31] Rather, glare resulted from misjudgment on the part of an untrained public. Brighter lamps looked to many like the very emblem of technological progress, leading people to install lamps that were simply too powerful.[32] Architects could contribute to the problem by specifying lighting fixtures originally designed for candles but, in deference to client wishes, use electric bulbs.[33] To be dazzled was precisely why people went to electrical exhibits at expositions or stared at new street lamps, surmised a Chicago-based electrical engineer and patent attorney. They wanted the visual experience of excessive brightness, he wrote in 1895: "They wanted to have this intensity kept prominent—wanted to be dazzled by the penetrating rays even if they were thereby kept from seeing what was supposed to be lit up by the lamps."[34] That is, they sought glare as an unprecedented and thrilling visual experience; they longed to surrender to the luminous spectacle as if optical discomfort were itself a register of material progress. Across a range of settings, the frequent mismatches between the methods of lighting and the needs of vision came at this time to be called "the glare problem," understood generally, in a commonly repeated contemporaneous phrase, as "light out of place."

As much as it was an inconvenience or, in a number of situations, a hazard, glare was also an ideological problem. To the extent that light was an emblem of intellectual advance and electric light in particular a material affirmation of that advance, glare introduced discord, even contradiction. Glare was a false conclusion derived from an unimpeachable syllogism; it was sand in the gears of progress. It made the dual character of technological progress—that it presented perils along with its conveniences—painfully obvious. It smuggled an ephemeral visual infirmity under the cover of improved illumination and, in so doing, turned attention from the objects of vision to the mechanism of vision, eliciting a sense of the struggle to see and revealing a precarious, almost fragile aspect of human vision, even (as metaphor would have it) human reason. As a by-product of increased electric lighting, glare moved in the same direction as technological progress, but it gave rise to an effect that was the opposite of what was intended. That is, the glare of electric lighting created an obscurity, a kind of conceptual darkness that electric light was supposed to dispel.

In *Madness and Civilization,* Michel Foucault notes that "truth and light, in their fundamental relation, constitute classical reason." Since antiquity, the "circle of day and night" has constituted "the most inevitable but the simplest of nature's legalities." But that clear distinction is at the same time pure deception. In madness, and even in ordinary circumstances, the two intertwine. Madness, Foucault points out, is an equinox between nighttime hallucinations and daylight's reasoned judgment. To be mad is to be "drunk on a light which is darkness," or to follow false lights. Following the optical metaphor of glare, Foucault asserts that reason may become so consumed by brightness as to become blinded, to the point that it "sees *nothing,* that is, *does not see.*" Thus, he concludes, madness must not be understood "as reason diseased, or as reason lost or alienated, but quite simply as *reason dazzled.* Dazzlement is night in broad daylight, the darkness that rules at the very heart of what is excessive in light's radiance." Individual objects of vision "disappear into the secret night of light, sight sees itself in the moment of its disappearance." In madness, delirium and dazzlement displace truth and light.[35] In this context, glare, a new aspect of nocturnal perception, was not only a bugaboo of electric light. It was an optical nullification at the heart of technological progress.

STRUCTURE OF THE BOOK

Today, after more than a century of electric technology,
we have extended our central nervous system itself
in a global embrace, abolishing both space and time as far
as our planet is concerned.
Marshall McLuhan, 1964[36]

The following chapters present five examples of ways in which electric light created new spaces or so substantially altered visual conditions in existing milieus as to compel new behavioral protocols or new codes of social relations. Electric light's distinctive traits—its advantages over prior forms of lighting, its relative brightness, its portability, its malleability and combinatorial possibilities, its instantaneity and action at a distance, its introduction of glare, and its eventual naturalization into the commonplace—are interwoven and recur thematically throughout these case studies. Electric light is presented as a singular engine of modern experience that is embodied in plural forms and to various ends. Although differing along multiple dimensions, the cases are organized more or less in increasing scale, from the small, tangible light switch to the immaterial and borderless gloom of a blackout. The general time frame covered runs from the late nineteenth century through the 1940s, that is, from the introduction of electric lighting in textile mills to the visual anxieties precipitated by blackouts during World War II, which demonstrate the degree to which electric lighting had become an essential and ubiquitous element of everyday life. Most of the discussion concentrates on developments and perceptions between the 1890s and the 1920s, the period of time when electric light went from being novel to normal.

Chapter 2, "At the Flip of a Switch," sketches the history of one of the most startling and personally empowering actions at the start of the twentieth century: switching on an electric light. For the first time in history, an ordinary person could at will alter the appearance of a room instantly and from a distance. With minimal effort, a room would rocket into visibility. To switch on an electric light was to suggest, however inchoately, a realm where anyone ruled like a god, capable of willing worlds into presence through nothing more than stretching out a finger and pressing a button. The power of such a transformation was not lost on politicians, who made flipping the switch on public works projects a staple of political theater.

Chapter 3, "Driving through the American Night," treats the roving cone of an automobile headlight as a new type of space that was unprecedented in its ubiquitous mobility. Suddenly in conflict with other spaces and other perceptual habits, the illuminated cone produced by the headlight became distinctly social: it was designed, regulated, adjudicated, and gendered; it mattered greatly whether you found yourself inside it or outside, whether you followed it from behind or saw its leading edge rapidly approach. Drivers and passengers sat at the vertex of a luminous cavity, pursuing but never entering it, all the while moving through a mantle of dark; at the periphery other drivers and pedestrians were insulted by the visual violence of sudden glare and the peril of collision. While comparable in many ways to cinema, a coeval invention, the visual field generated in night driving is the product of the body's mechanically driven motion through space rather than the virtual recreation of motion apparent in cinema. However acclimatized our routine encounters with the headlight have since become, the story of the headlight's rapid diffusion—shaped, navigated, avoided, adjudged, accommodated—is an extraordinary episode in electric light's creating and sowing a new space.

Chapter 4, "Lighting for Labor," considers factories as they came to be lit by electricity. Dedicated spaces for labor and its illumination have long been a concern since the quality, speed, and duration of productive activities are affected by the way in which they are lit. But the degree of control possible with electricity heralded a new level of precision in lighting. A host of experts arose, including illuminating engineers, physiologists, efficiency experts, and industrial managers, all concerned to devise new scales not only to measure lighting levels but also to quantify its contribution to overall output. For many, lighting objects was a lesser issue than enhancing workers' eyesight as a means of increasing productivity.

At a moment when ideas of scientific management flourished, such experts extended labor relations into the field of perception, effectively trying to Taylorize vision. Seeing, in essence, was being commodified.

Chapter 5, "Electric Speech in the City," looks at Times Square, New York, to consider the formation of a new type of urban space: the zone of illuminated commercial speech. Here, electric light was the vanguard, means, goal, and emblem of the colonization of the public sphere by private interests. A representation of consumer capitalism was founded on light, becoming a new urban district and touristic destination. It began with the "Great White Way," a stretch of lighting and commerce that gradually moved up Broadway to settle and swell in Times Square as a defined district dedicated to lighted commercial speech and heightened visual stimulation. Soon, nearly every ambitious American city clamored to be judged by the brightness of its own White Way, seeing it as a self-evident sign of progress and, it hoped, a spur for tourism and population growth. At the same time, cities throughout the world established similar districts, often comparing them to Times Square, which thus became an engine for the globalization of a particular form of urban culture.

Chapter 6, "Groping in the Dark," begins in the 1930s with electric light as one of few topoi of continued material progress during the Great Depression. Government-mandated wartime blackouts in response to rising global hostilities and uncertainty regarding the possibility of air raids over North America came as a tremendous surprise and precipitated a crisis in the visual field for many Americans. Countless individuals and groups began to consider life in scotopic space, or space as perceived under conditions of little or low lighting. Widespread ambivalence regarding whether or not blackouts were needed in the first place prompted numerous explicit justifications from a remarkably wide range of advocates. Just about everyone—from government experts to medical researchers, civil defense specialists, soldiers and pilots, judges and attorneys, investment advisers, and poets, not to mention ordinary citizens—had advice regarding ways to adjust to dimmer surroundings, to infer spatial information from nonvisual senses, to familiarize oneself with nightscapes based on reflected rather than geometric properties of surfaces, and generally how to inhabit and navigate a darkened world. Although the blackouts lasted only a few years in the United States, and no city was ever bombed, they reveal the profile, though in negative terms, of how the nation's visual environment was imagined around 1940, sixty years after electric lighting began to change the way we understand and inhabit space.

2
AT THE FLIP OF A SWITCH

> If I stood on the bowbacked chair, I could reach
> the light switch. They let me and they watched me,
> A touch of the little pip would work the magic.
>
> **Seamus Heaney, "Electric Light," 2001**[1]

INTRODUCTION

At midnight on December 31, 1965, Pope Paul VI pressed a button in his apartments in the Vatican to switch on the lights illuminating the statue of Cristo Redentor on Mount Corcovado, in Rio de Janeiro, Brazil, some 6,000 miles away (figure 2.1). Begun in 1922 to celebrate Brazil's centenary, the statue stood on public land, underscoring the Catholic Church's prominent role in the nation's political identity.[2] The new lighting installation, planned by the prominent American lighting designer Richard Kelly, was part of the city's 400th anniversary celebrations, which included a year of parades and festivities and, on the last day of the year, thousands of white candles and flowers along the beach, women dressed in white, and copious amounts of the "white" alcohol cachaça. Having the pope illuminate the mountaintop statue from his seat in the Vatican was seen as part of a typically Cariocan mix of the religious and the political or, as one newspaper put it, progress and paganism, and of consecration and commerce:

> At the stroke of 12 (Rio time)—Pope Paul VI touched a button in the Vatican and illuminated, with a new set of floodlamps, the figure of Christ the Redeemer that overlooks the city from atop Corcovado Mountain. Air Force planes flew overhead, dropping "silver rain" of bits of foil painted with the name of the State Bank.[3]

It is unlikely that the pope's button directly controlled the electricity flowing to the lights at the base of the statue. More likely, he sent a signal to an operator situated adjacent to the light switches. But by touching a button, the pope completed a virtual circuit that looped his own body with the mountaintop figure of Christ half a world away. The gesture reaffirmed—indeed, embodied—the connection between the Brazilian state and the Catholic Church. But it did so in distinctly electrical terms: the smallest of human gestures, a touch, summoned a force of nature, instantly and across a great distance.

In making a figure of God visible to man, the pope, along with the press, relied on several characteristics of electric light that resonate with divine action. First, it can be exercised from far, far away—the pope was on the other side of the Earth when he ignited the lamps. Second, it is effortless—the pope had only to touch a button (God, presumably, does not struggle in the creation of earthly events). Third, it is instantaneous—the pope's will, his gesture, the visual effect, and the crowd's awe were virtually simultaneous. Fourth, like God, electricity is unseen, known only through its effects, and historically has been most visible as one form or another of radiant light. As such, electric light participates in an ancient religious iconography, despite its technological provenance.⁴ Light is laden with literary and pictorial tropes heavy with scriptural significance and the subject of voluminous investigations: at many points in Western religious history it has indexed the very presence of God. Rather than address such momentous matters, however, the following discussion looks in a different direction, toward the button.

Although small and easy to overlook, the button, or switch, is a highly charged interface between individuals and technological systems. Turning a light on, instantly and at a distance from the lamp, at first felt like a kind of magic, especially since electricity was so unlike other fuels or forces or sources of light. With a switch, anyone could change the space around him- or herself, visually at least, on a whim. It put the resources of what soon became a vast infrastructure at hand, literally. A common trope held that the pressing of an electric button was an easier way to summon the "modern genie" of great power than Aladdin rubbing his lamp.⁵ Politicians seized on the switch's powers of instantaneity and action at a distance to inaugurate a new way to celebrate public works projects: the switch became a live-action metaphor of the authority and ability to get things done. With these uses, the switch generated a visual space whose merit was measured best by the gap between its effortless operation and its broad scope of effect. Its symbolic power rested on a disproportion: the minimum haptic gesture sparked the maximum optical transformation.

2.1
Cristo Redentor, illuminated with new lighting equipment on December 31, 1965, by Pope Paul VI from the Vatican. *Cristo Redentor*, photograph by Dia Mundial do Diabetes, CC BY-NC-SA 2.0

A MATERIAL CULTURE OF CONTROLS

Another point of difference in which gas cannot possibly compete, is the manner in which the electric light can be turned on and off, and which admits of the switch being placed at any convenient point. This is denied to gas.
Edward De Segundo, 1893[6]

The light switch is part of a long history of control mechanisms that regulate an otherwise continuous flow. In this sense, it is antique in conception, analogous to dams or sluices or gates that control a flow of water. Although there is no evidence for valves in ancient oil lamps, the imperial Roman Egyptian author Hero of Alexandria described something akin to an automatic fuel supply that worked by means of a hydraulic or pneumatic contrivance.[7] A window shade or shutter might also be considered a type of switch that controls the flow of visible light. Similarly, a door controls the movement of matter between inside and out or between one room and another. Indeed, the English word "valve" is derived from the Latin term for the leaves of a double door. Starting in the eighteenth century, "valve" came to be used in English in relation to water control. The analogy between valves and electrical switches continued at least through the nineteenth century. In an 1892 patent infringement case brought by the British electrical engineer St. George Lane-Fox, the judge, for whom electrical switches were still a new idea, learned during the proceedings that "switches were as commonly used in electricity as water taps in the distribution of water."[8] Electricity, an invisible force whose operation lay outside ordinary experience, was often explained by means of a hydraulic analogy employing terms such as hydraulic head, flow rate, and pipe diameter to help students grasp electrical potential, current, and resistance.[9] The analogy, although inexact, remains common today.

The English term "switch" derives from a riding switch, a long stick used to indicate to a horse the rider's interest in greater speed, essentially to convey a rider's will wordlessly and convincingly. Railway operators picked up the term around the 1820s to designate the set of rails used to shunt trains from one track to another; the lever an operator pulled to effect such a shift was called a "switch-rod." Generalized, the switch was a means to change the configuration of a track or circuit, which facilitated its later application to electrical circuits. Its use in the mid-nineteenth century to describe a telegraph key drew on several of these earlier senses: as a slender stick, a cause to initiate or alter action, or a means to complete a circuit. In relation to the telegraph and its unprecedented speed of transmission, as if thought itself leapt across continents, the switch was seen to inaugurate a compression of time and a collapse of space. By the end of the century, "switch" denoted a manipulable tool used to regulate potentialities within a larger current of objects or forces, most notably electricity.[10]

Earlier forms of lighting were not so much switched on as they were prepared for illumination. Generating light from an oil lamp, for instance, required several steps—removing a protective enclosure for the flame; assuring the fuel supply; examining and adjusting the wick; igniting it (no small task in the days of the tinder box, prior to the commercialization of wooden friction matches); adjusting the flame; and, finally, lowering the enclosure.[11] These steps were sufficiently distributed in character, proximity, sequence, and requisite dexterity to preclude use of the term "switch," which bears the suggestion of instantaneous operation. Although gas lamps were simpler to light than oil lamps, they still required a sequence of steps that had to be repeated in turn for each lamp.[12] Extinguishing the flame was easier since it required only closing the valve supplying the flammable gas. Theaters pioneered efforts to develop systems to light multiple gas lamps remotely, sometimes with flames at the end of long poles, but these often proved unreliable, with the unhappy consequence that theaters were frequently incinerated in the nineteenth century.

While valves had become relatively familiar as cities were woven through with water and gas, electric switches started becoming common only in the late nineteenth century. As the electrical industry grew, switches began to appear across a range of new commercial products such as irons, fans, toasters, kettles, and home sewing machines. After visiting an electrical exhibition held in Philadelphia in 1884, one reviewer, for example, noted, "The alternate making and breaking of an electrical circuit is an old device, but its recent applications have brought it into the domain of marketable commodities."[13] Electrical switches were even then still a novelty, and unthinkingly making and breaking circuits was a habit yet to be acquired. For those who could afford electricity in their own homes at that time, lights, more than anything else, were the primary appliance getting switched on and off.[14]

Engineers and inventors designed many configurations of switches to account for a range of situations. Switches had to tolerate line voltages that were more variable than they are today, and they might have to accommodate higher loads over time as more devices were added to existing circuits. Engineers also had to consider the location of a switch in relation to the heat it gave off or to the consequences of failure. Even variations in users' dexterity could be a factor. For instance, with a knife switch, a hinged lever that was raised and lowered into a metal slot, slow operation could generate sparks or electrical arcs and thereby create a fire hazard. Early switches could be little more than a "circuit-closing screw" operated directly at the fixture, as described in an 1880 patent filed by Thomas Edison, which required someone to twist a contact screw one way to turn on the light and the opposite way to turn it off, similar to a valve in a gas lamp (figure 2.2). The circuit "is either completed or broken, dependent upon the direction of the turning." Partial turns, the patent claimed, would result in variable light-

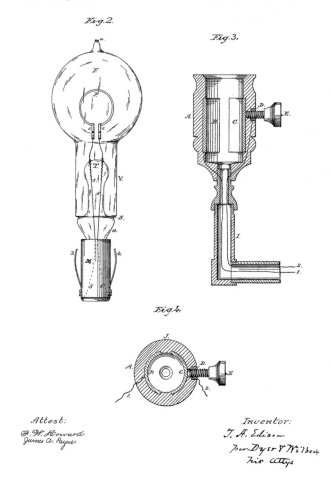

2.2

Edison's electric lamp holder, operated with a metallic thumbscrew, is part of a lamp "complete in itself." The design, while feasible, risked sparking because of unstable voltages and, under certain conditions, could have produced a shock to the user. Edison, Thomas Alva Edison, System of Electric lighting, U.S. Patent 265,311, filed February 5, 1880, and issued October 3, 1882.

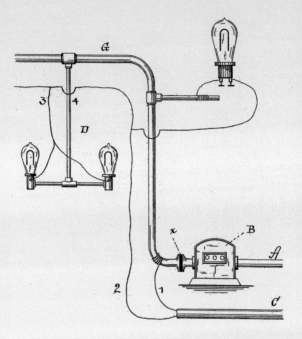

2.3
Diagram illustrating the use of existing gas piping as conductors to create an electrical lighting circuit. No switch is described as part of the system. Thomas Alva Edison, System of Electric Lighting, U.S. Patent 251,551, filed August 30, 1881, and issued December 27, 1881.

ing levels, from "the total lighting effect, a dim lighting effect, or no lighting effect being due respectively to a fine contact, a slight contact, or no contact" between the screw and a conductive plate.[15]

The design of a switch, even its necessity, was not immediately obvious in consumer settings. Another Edison patent, from 1881, used gas piping as conduit and included such details as bulb fittings and insulation from the gas meter. But it made no mention of a switch within what was supposed to be a complete system of electric lighting for home use (figure 2.3).[16] In just a few years, however, manufacturers were producing a wide range of light switches with various types of actuators, or controls, whether toggle, rocker, push-button, rotary, and so on. The convenience of the switch was so great that manufacturers even began to offer electrical switches for gas lamps that could open a valve and ignite the gas from a remote plate on the wall, thereby acquiring "one of the great advantages of the incandescent electric light" that was just then threatening to displace gas altogether (figure 2.4).[17]

FIG. 2.—FIXTURE WITH "UNIVERSAL" SWITCH.

2.4
Diagram intended to illustrate how a switch may be minimized so as not to detract from the style of any lighting fixture. "The 'Universal' Electric Light Fixture Switch," *Electrical Engineer* 10, no. 113 (July 9, 1890): 37.

VOLITIONAL SPACE

We have chained a giant that we do not know ...
at the press of a button he answers to our call.
W. J. Keenan and James Riley, 1897[18]

The switch was more than a practical and convenient means of turning lights on and off. In the early days of electrification it was also an assurance that electricity, a potentially lethal force previously well known only in the form of lightning, was easily controlled. The physics of electricity lay outside the grasp of the public, which heard horrifying descriptions of accidental electrocutions, trembled at the thought of planned electrocutions, or read expert testimony regarding the electrical industry's intensification of the natural atmospheric charge, which, some believed, could accumulate and one day explode. Court cases that featured electrical mishaps helped garner public support to regulate the industry.[19] In 1896, after more than 300 pages that attempted to answer the recurring question, "What is electricity?," a Harvard physics professor came to the simple conclusion, "*Ignoramus ignorabimus*—(We are ignorant, and we shall remain ignorant)," before unhelpfully pointing out that the Earth is already bathed in electromagnetic waves from the Sun.[20] Even into the twentieth century consumers worried about having wires in their home, fearful of shock or concerned that electricity could leak from outlets, as James Thurber humorously recalled his grandmother believed.[21] Thus the first explanations of electricity for a general audience, from the 1880s, emphasized its "complete amenability to control" by means of well-placed switches.[22] The surest testament to both the safety and ease of use of electric lights was the claim that even a child could operate them safely, a trope employed in the promotion of a variety of new technologies:

> If you will watch me going the round of this room, you will see how unscientific are the means used to turn on and off the lights. One has simply to twist the switch near the door to put out the whole of the lights in the room. ...
>
> Can it really be said, after this, that the electric light does not fulfill the condition of being under perfect control? Why, my little boy, of seven years old, knows as much about this part of the subject as I do, and, with a switch attached to his bedside, can turn on the light in his bed-room whenever he requires it.[23]

Despite its diminutive appearance, the switch guaranteed "perfect control," evident in its easy operation by a diminutive person (figure 2.5). Whatever its technical configuration, the switch domesticated electricity, tempering it sufficiently to bring it inside the home and make it serve at the pleasure of even the most innocent occupant. It confirmed that ordinary lived space, collapsed and reconfigured by modernization, could nevertheless be made whole and amenable. It promised greater personal control even as it expedited and extended electricity's entry into intimate realms such as bedrooms and, in so doing, quickened the pace of modern life to the twist, flip, press, or snap of a switch. At a moment when modern inventions such as photography and practices such as the increasing circulation of magazines and newspapers encroached on accepted notions of individual propriety sufficiently to inspire Samuel Warren and Louis Brandeis's 1890 defense of a right to privacy, the switch recoded modernization as a triumph of individual will over sophisticated and corporate technologies and practices rather than a disintegration of individual sovereignty. In this way, the switch acquired an emotional and psychic register.

2.5

Unexpected surprise: a mischievous boy switches on the lights, catching his parents in an amatory moment. The scenario suggests both the switch's ease of operation (it can be manipulated by a child) and its capacity for erotic revelation. The illustration is part of the Light Is the Life of the Home series of advertisements prepared by Norman Rockwell for General Electric and appeared in the June 12, 1920, issue of the *Saturday Evening Post*, p. 61.

Exercising control over threats in one's environment, as Sigmund Freud suggested, is an early developmental challenge. He recalled his eighteen-month-old grandson throwing objects away from himself, saying what Freud heard as the German word *fort*, meaning "gone" or "away," and then, in retrieving the object, saying *da*, or "there." Freud theorized that the child was symbolically replaying his mother's temporary absence, which, however painful, was at the same time prelude to her return, a source of joy for the boy. Among possible rationales Freud offered for this behavior was that the child was expressing "an instinct for mastery" of his world that required him to shift from being a passive witness to being an active agent in the shaping of events, at least on a symbolic level. The boy mastered the loss of his mother and rehearsed her return, a game that was validated by her eventual reappearance.[24] The regularity of the toy's return suggested the reliability of the continuing cycle of presence, which brought pleasure, and absence, which, if unpleasant, could nonetheless be endured.

Electric lights do not figure in Freud's explanation, but it takes only a small step to liken the two poles of the *fort-da* game, not-here and here, to the off and on of the switch, especially in the face of anxieties regarding powerlessness in an increasingly complex environment. The underlying mechanism is similar both for its pattern of repetition and for the action of a will to resolve and regulate the alternation of two opposed states. Switches are clearly both implements and symbols of mastery. As the electric light came to be aligned with psychic processes, it was likewise seen as a symbolic means of asserting control over a situation. Pressing the button, then, was not only a way to control a new and threatening technology. It represented as well a kind of self-control and a reassurance that the threats posed by the penetration of technology into everyday life—the wires woven behind walls; the switches and fixtures bristling from surfaces—would not similarly penetrate the self and disrupt individual integrity.

Freud's interpretation of his grandson's psychic development may be challenged, but his theory offers a model for understanding the role of the light switch in the formation of a modern sense of the lived environment. Far from being a mere void or neutral entity into which objects are placed and through which systems are threaded, the space generated by the switch was an *activated space*. It was a space organized to enhance visual perception and animated by its guaranteed responsiveness to an individual's will. For this reason, the space illuminated by the switch can be deemed a *volitional space*, a volume summoned into visual presence by the sheer desire to see. A mastery of self can be understood to entail a mastery of space or, perhaps more precisely, a mastery of the self-in-space, wherein self-control is predicated on spatial control. Operating the switch, then, rehearses the sense of self-control that is threatened by transfiguring technological systems. Each repetition of the switch's off-on cycle promises that another cycle will surely follow: more technological mediation of everyday life coupled with easier and more reliable control.

The switch induced a new and modern space defined not by size, shape, structure, material, use, ornament, or any other conventional measure of architectural merit. Rather, it conjured a space distinguished by its instantaneous appearance, a volume willed into visibility, as if volition alone were sufficient to make it so. Indeed, the idea of a volitional space presumes that individual will is as much a part of the visual transformation created by electric light as the switch mechanism's metal contacts. Visibly projecting willpower into a third dimension, volitional space is as much the amalgam of technology and desire as it is an image of desire reliably fulfilled.[25] A light switch is then best understood not simply as a mechanical device but as an alloy of volition, body, and mechanism.

In earlier spectacles, such as moving panoramas unrolled before an audience, or in a theater more generally, the viewer was passive. The switch, however, allowed the viewer to produce the spectacle; the venue, in turn, was infused with the individual's will. Perhaps the most remarkable example of this is what has come to be called "path lighting," wherein lights can be switched on and off so that one moves through an otherwise darkened building within an envelope of light. Engineers developed easy-to-use three-way switches that allowed lights to be switched on or off in sequence. As one lighting advocate explained to homeowners, "The householder should be able to visit the entire building, commencing with the hall door, from attic to cellar and back, without once being left in the dark, or leaving lamps burning on any floor behind him as he makes the journey."[26] Facilitating a journey through domestic space in a vessel of light operated by a series of buttons, electric path lighting recapitulated by modern means an older method of navigating dark spaces by holding up a candle.

The light switch, to put it another way, is a modern prosthetic, as much a material extension of the will to see as it is an instrument of control (figure 2.6). As new technologies such as electricity spread through cities in the early twentieth century, the tiny drama of homeowners switching their new electric lights on and off rehearsed and thereby reinstated a sense of agency otherwise diminished by the infiltration into the fabric of everyday life of poorly understood power networks, which required specialized trades to maintain (figure 2.7). In sum, the switch did not just turn lights on and off instantly and at a distance. It also brought about a simple but impressive new reality: for the first time in the history of humankind, ordinary people could alter the visual appearance of their living quarters at will and instantly. From there followed a host of other possibilities.

2.6

It's on, it's off: the function and narrative of the switch. The illustration appeared as part of an advertising campaign for electrical systems for farms, from a manufacturer of agricultural equipment, Rumely Products Co., of La Porte, Indiana, around 1912–1913. Advertising Ephemera Collection, John W. Hartman Center for Sales, Advertising and Marketing History, David M. Rubenstein Rare Book and Manuscript Library, Duke University, Durham, North Carolina.

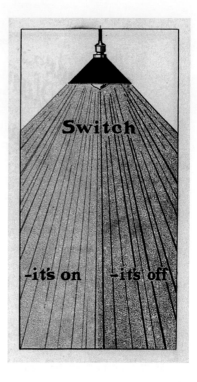

2.7

The artist Walter Crane, on visiting Chicago in 1893, stayed in the hotel in Louis Sullivan's Auditorium Building. He toured skyscrapers with push-button elevators, observed suburban women using the telephone to order groceries, and visited the World's Columbian Exposition while it was still under construction. After these exposures he was inspired to draw *The Button-Presser*, a prophecy of human evolution. Walter Crane, *An Artist's Reminiscences* (New York: Macmillan, 1907), 379.

THE BUTTON-PRESSER—FANCY PORTRAIT OF THE MAN OF THE FUTURE

THE AGE OF BUTTONS

As soon as it is dark enough to need the artificial light,
you turn the thumbscrew and the light is there.
New York Times, 1882[27]

Is not this somewhat the way that God works?
The Independent, 1893[28]

The President's Finger

With its ready metaphor of electrical energy couched as volitional power, the switch played a starring role in a new form of political theater, one that continues to the present day. The actors were varied, ranging from business executives and mayors to presidents and popes. But the plot was always the same: a crowd assembled at dusk, often to listen in the gloaming to a speech about progress; a button was pushed and a lighted landscape abruptly loomed, as if the throng's will were concentrated in the fingertip of its political representative and transmuted in a flash into radiant accomplishment. Crowds typically greeted the visual transformation with an audible gasp. Whatever needed inaugurating, a politician's finger could reliably be found inching its way across the podium toward a switch that would suddenly reveal a new and modern triumph.

Millions saw their first incandescent lights switched on at expositions. From Philadelphia in 1876 to Paris in 1937, electric lighting was an explicit theme at such events and a leading fairground attraction. In nearly every instance, switching on the lights was an entertainment in itself. On May 1, 1893, at Chicago's World's Columbian Exposition, President Grover Cleveland pressed a lever to start the dynamos that would provide electricity to the vast array of equipment, from large engines to powerful arc lights, scattered across the site. He surrounded himself with descendants of Christopher Columbus's family, as well as his vice president and members of his cabinet, to vivify the historical narrative that was at stake. The lever itself was wired to a temporary circuit hooked up between the switch and a single-cell dry battery, which in turn was wired to the enormous Corliss engine standing in Machinery Hall, ready to unleash its energy.[29] A reported 100,000 visitors showed up to witness the president's finger as it hovered over and then dropped down to touch a button mounted on the lever's end (figure 2.8).

The pedestal supporting the lever, golden in some reports, was "a pyramid of blue, gold and plush 12 inches high" and intended to represent America and Spain. A small plaque at the base of the pyramid bore the dates 1492 and 1893, making clear that with the closing of the electrical circuit at the exposition, a 401-year historical circuit would also be completed, the United States having evolved during that time to become a leader among nations.[30] Cleveland's speech likened the

2.8
Thousands assembled in Chicago in 1893 to see the president press a button to officially open the World's Columbian Exposition. The event was reported internationally. Rudolf Cronau, "Weltausstellungsbriefe aus Chicago," *Die Neue Gartenlaube* 13 (1893): 401.

instantaneous effect of the switch to the achievement of American ideals. He punctuated his conclusion by raising his finger over the button as if his words, conveyed by the medium of his touch, might themselves turn the wheels of progress. In Cleveland's view, the switch operated several circuits: the historical one, a track of virtuous American power circling the globe, and a loop of copper wire running through the fairgrounds. "As he pronounced the last word he touched the ivory button of the key of gold on the desk before him, and the spark of electricity flew to do the President's bidding."[31] It was surely a stirring moment, not least for Cleveland, from whose simple gesture an entire world was set in motion, as a contemporary account put it: "When he touches the button every wheel starts, every process of beautiful production goes on before our eyes."[32] The president's words, his fingertip, and a force of nature were united, vehicles all for American will.

The meaning of the moment was lost on no one. Pressing a button in public placed the button presser at a potent threshold between worlds. As banal as it was—the gold plating and ivory buttons only attest to its underlying inconspicuousness—the switch was lodged in an intersection between authority and the territory it oversaw. Button pressing shed reflected light on a single figure, or, to put a point on it, a single figure's finger. With a frightening force of nature bent to beam light and start great engines, the button presser appeared Olympian. A newspaper editorial made the point plain just after President Cleveland switched on the power in Chicago:

> Deeds of grandeur or deeds of terror are accomplished with less immediate effort and at a distance from their effect. The touch of a button executes a murderer or starts all the enginery of the Columbian Exposition. Is not this somewhat the way that God works? ... No cause appears; but suddenly he dies. ... The touch of a button by the President starts into active motion the ponderous machinery of the Exposition. Where was he? Invisible, somewhere else. ... Where was He whose will created and set in motion the processes of nature ...? We do not see him; perhaps we forget him; but had we looked we should have found his finger at the keyboard of the universe.[33]

The measured drift between President Cleveland and God made the message crystalline. On the one hand, Creation itself was as simple to God as pressing a button, and on the other hand, humankind, or in the context of that day the United States, now possessed by dint of technological mastery a kind of divine agency and could administer its collective will as easily and decisively "as a thunderbolt leaps from the sky."[34] Projecting one's will across space and bidding an otherwise invisible energy into material presence rehearsed a mythic moment of creation and echoed something of the divine: whoever pressed the button commanded power as would a god. Indeed, the switch was a paradigm of divine power: local in operation, universal in scope, easy to wield, awesome in effect. The elaborate ceremony, the ponderous speech, and the overwrought switch testified to something everyone clearly understood: the meaning of the switch lay in a gesture, not a thing, in a minimal action rather than an ornate artifact. The switch's imaginative power resided in the pivot between its complete banality and its outsized effect.

Buttons lightly pressed turned into a leitmotif of the exposition, as switching on the lights became a nightly entertainment. Charles Edward Bolton, a turn-of-the-century travel writer and the husband of temperance leader Sarah Knowles Bolton, recounted the crowds that gathered nightly to witness "the silent touch of an unseen hand" that illuminated the bulb-spangled Edison Tower and a host of other electrical lights and devices: "Other buttons touched, revealed throughout the hall the splendid workings of mind with that subtle something, which is called electricity, till the whole became a marvelous fairy-land." (figure 2.9)[35] Still other buttons touched allowed fairgoers to take pictures with the Eastman Kodak instant camera, introduced only five years before. To underscore its simplicity, the company launched one of the most famous advertising campaigns in American history: "You press the button, we do the rest."[36] The idea of an easy exercise of mastery over technical processes merely by pressing a button was said to be "really the prophetic cry of the age."[37] In both cases—of the electric light and the portable camera—tiny taps sparked visual revolutions (figure 2.10).

2.9
By the time the 1893 World's Columbian Exposition was over, millions had witnessed the lights being switched on at the Edison Tower in the Electricity Building. John Patrick Barrett, *Electricity at the Columbian Exposition* (Chicago: R. R. Donnelley and Sons, 1894), 17.

2.10
Switching on the lights, and the switches themselves, remained for years a source of fascination at expositions. The "one little switch that turns out all the lights" is barely visible at the far left of the image. *In the Electric Building*, in Holman F. Day, Arthur Hewitt, photographer. "Three Pilgrims at the 'Pan,'" *Everybody's Magazine* 5, no. 26 (October 1901): 436.

AT THE FLIP OF A SWITCH

Political Theater

In other words, let us "turn on the light!"
Franklin Roosevelt, 1932[38]

Switches became a staple of political performances. Lighting Christmas trees from afar was an annual favorite. In 1903 Theodore Roosevelt pressed a button to start the holiday season, as would Woodrow Wilson, Calvin Coolidge, and Herbert Hoover in their turn. (figure 2.11)[39] Central to all accounts was "the simple action of throwing an electric switch," the momentary mechanical act that spread a season of joy and camaraderie across the land.[40] Mayors too joined in, sometimes deputizing their own or others' children for the occasion.[41] In countless towns and cities, switching on the lights was the primary way to celebrate the *ability* to switch on the lights following the installation of municipal lighting systems.[42] Even popes were drawn to the tiny drama of pressing a button as a way to demonstrate jurisdiction and to recall the role of light as a metaphor of divine creation. Pope Pius XI, for example, noting that the Church's mission on Earth would materially benefit from electric light, switched on the new dynamos that would illuminate the Vatican in 1931.[43]

2.11
President Harry Truman made a grand show in 1945 of switching on the lights on a Christmas tree. Coming at the end of World War II, with memories of government-mandated blackouts still fresh, the gesture was especially meaningful to Americans. *President Truman Lighting a Christmas Tree,* December 24, 1945. Harry S. Truman Presidential Library and Museum, Independence, Missouri.

Early in the twentieth century, industry trade journals, along with many popular sources, often reported on local luminaries who pressed buttons to turn on new lights. In many cases they clearly appreciated the reflexivity of the gesture. The pageantry of switching-on ceremonies and the radiant lighting up of streets formerly dim demonstrated that "electric service is regarded as something more than a mere commodity," especially "by those who have thus far been forced to get along without it," as one commentary ran in 1922. This was followed by a number of notices regarding lighting celebrations: a dream come true in Beaver Falls, Pennsylvania; a button pressed in Rockland, Maine, followed by a parade with a prancing charger horse, a man in a pumpkin-head costume, and the local Elks playing brass horns; Roscoe (now Sun Valley), Los Angeles, "celebrating under the incandescents," with a festival of "dance and appropriate exercises"; and more dancing in the streets, country music, and "jazz clamor" in Montebello, California, where men were invited to "Come on over, and bring the girl along" to see the lights go on.[44]

As the power behind two ambitious electrification initiatives, the Rural Electrification Administration (REA) and Boulder Dam, Franklin Roosevelt was an enthusiastic switch flipper, often turning up to launch new lighting systems by publicly throwing a switch. The performance usually involved a fake switch set up on a podium. The actual switch would be some distance away with the operator listening by radio and cued to a key point in Roosevelt's speech. At least once, in a contentious speech endorsing the opponent of a sitting senator of his own party, the president, impassioned, went off-script, forgetting to signal the operator and leaving the lamps unlit until some minutes after the speech ended.[45]

The most powerful button Roosevelt pressed was the one that sparked the turbines of Boulder Dam into action, in 1936. The dam was crucial for Roosevelt because it was expected to fulfill his promise to lower the cost of electricity. More available power would spur higher demand, which would lead to more power—a virtuous circle of economic growth. Boulder Dam, on the Colorado River near Las Vegas, was only one of a number of dams built to generate electricity, but it would produce the greatest amount of energy ever generated, at the highest voltages ever achieved, and it would transmit the power farther from its source than any previous dam had—to Los Angeles, nearly 300 miles away. For his part, Roosevelt was in Washington, D.C., some 2,400 miles distant, when he pressed the button (figure 2.12).

The event, timed to coincide with the Third World Power Conference and the Second Congress on Large Dams, was attended by thousands, including 700 delegates from more than fifty countries. With radio microphones around him, Roosevelt spoke of the boon to mankind that followed electrification. At the end of his speech he raised his finger over a gold-plated button, intoning:

> At this moment the powerful turbines are awaiting the relatively tiny impulse of an electric current which will flow from the touch of my hand on the button which you see beside me on the desk, to stir machinery into life, to stir it into creative activity to generate power.

He pressed down and continued, shifting into a second-person incantation: "Boulder Dam! In the name of the people of the United States, to whom you are a symbol of greater things in the future; in the honored presence of guests from many nations, I call you to life!" After a short pause, the turbines began to turn.[46]

As a national grid began to take shape in the 1920s and 1930s, electricity was seen by many as a binding force. It was possible then to imagine national cohesion as at least partly an *effect* of electricity. Certainly, Roosevelt's REA suggested as much, with its goal of uniting country and city through the modern amenity of electrical power, regardless of geography. Rural Americans could talk with urban Americans on the telephone despite the distance and could enjoy the same modern conveniences, at the press of a button, as city dwellers. The switch proffered a vision of a modern material civilization knitted together by devices designed to seamlessly serve human needs. Like radio, which fostered a common national conversation, the switch suggested a unitary space, in which presidents could start engines across the country.[47] By widening the franchise of material well-being from one coast to the other, the switch performed progress (figure 2.13).

2.12

President Franklin Delano Roosevelt starts the flow of electricity from Boulder Dam in 1936 by pressing a button. Bettmann, Getty Images.

2.13

President Kennedy presses a golden key after delivering telephone remarks to open the 1962 Century 21 Exposition in Seattle, Washington, from Palm Beach, Florida. Cecil Stoughton, White House Photographs, John F. Kennedy Presidential Library and Museum, Boston.

THE MAGIC OF THE BUTTON

Where has the world seen such magic before? A man in a power house turns a switch and a home many miles away is lighted. The turn of another switch—and the streets of a whole city with millions of inhabitants burst into radiance. The turn of still another switch sends a flood of light under the earth into the tunnels of a city where trains roar under the same power of electricity. Again, the turn of a switch lights up hundreds of miles of country roads. As late as the Eighteenth Century any man who had declared that such a thing might be might have been prosecuted as a madman or as a practitioner to the "black art."
Francis Trevelyan Miller, 1915[48]

In the late nineteenth and early twentieth centuries, pressing the button was routinely presented, if not exactly experienced, as a kind of magic. By the slightest of gestures a grand intention was realized, immediately and heedless of distance.[49] By extension, the space generated by the switch shared these properties; its appearance brought with it a penumbra of magical traits. Like all magic tricks, however, electric light depended on careful preparation. In 1902 in San Francisco, for example, workers had been at their jobs for weeks installing lighting on several streets in anticipation of the Knights of Pythias convention. But, one report went, no one noticed them or the web of wires they wove overhead until the opening day: "What they had accomplished was scarcely apparent until tonight, when by the touch of a button the city, still in a transition stage of architectural progress, was transformed into a veritable fairyland."[50] Only at that moment of metamorphosis did residents take note. As the report continued, the "effect of tens of thousands of artistically arranged incandescent globes ... springing suddenly into light, was marvelous. It was a realization of an Arabian Night's dream, or rather of the possibilities of twentieth century science."[51] The rational explanation for the effect, science, was submerged beneath a spell of wonder. Time, effort, and funding were subsumed in a florescent instant as the button eclipsed labor with light. Even breathless accounts of President Cleveland's button pressing were at pains to point out that despite the instantaneity of the event, the fair was the work of 15,000 men over three years at a cost of $33 million.[52] Yet the ultimate outcome of so much calculated effort registered, at least rhetorically, as a modern magic. In many instances, the switch was dubbed the "magic button." The power of the switch to render work as wonder flourished in this magical moment, emptied of time in its instantaneity and undaunted by distance in its remote efficacy (figure 2.14).

By seemingly transforming labor into light, the switch accentuated the commodious benefit of a public works project over the role of the workforce and supporting technologies. In this sense, the switch typifies the "device paradigm," an idea introduced in 1984 by the philosopher of modern technology, Albert Borgmann. The term describes a common trait of modern technology, that is, a particular configuration of components that eclipses a productive mechanism in favor of the commodity it delivers. Borgmann saw social relations in modern society as structured by the pairing of productive apparatus and a delivered commodity in such a way that consumption appears to be unmediated. Pipes and ducts, for example, separate the combustion of fuel from the resultant heat. They convey warmth while concealing the means of making it.[53] Though common to the widest range of tools, this cultural preference, facilitated by modern technologies, was neither inevitable nor neutral in its social effects. Indeed, Borgmann argued that dissociation from productive mechanisms nurtured ignorance of the social and material costs of commodities. As long as benefits were assured and costs predictable, consumers remained strangers to their own environments. The invisibility of the technology was proof of its effectiveness.[54]

2.14
"The Modern Aladdin," as much of the accompanying ad copy went, updated an old story of riches gained with a mere gesture. *Electrical Review* 69, no. 8 (Aug. 19, 1916): 321–322. Image courtesy of Warshaw Collection of Business Americana—Electricity, Archives Center, National Museum of American History, Smithsonian Institution.

2.15

Who Are You? Are you the sort of man who presses buttons and gets things done?, asks NELA, presuming, "You are the man whose second finger on the right hand is expert in pressing electric buttons." Remember, though, what power there is behind the button. Investors both large and small "make possible 'the button' that your finger presses morning, noon and night." With 14,000,000 homes still unwired, "foresighted citizens" stood to profit from higher demand for electrical service. Advertisement, National Electric Light Association, *Saturday Evening Post* 193, no. 34 (Feb. 19, 1921): 47.

At first glance, the switch seems to contradict the device paradigm since it refocused attention on the mechanisms of the electrical delivery system. But in experiential terms the switch suggested the ease with which vast amounts of labor, incalculable stores of energy, and sprawling networks could be snapped into service at a moment's notice. The switch not only controlled the flow of electricity, it represented that control. Its trifling size and trivial operation exemplified confidence in the ability to bring to heel a force of nature. Additional details, such as an ivory or gold-plated key, common adornments for ceremonial switching, or even an ordinary decorative switch plate removed the switch further from technical functionality and into the realm of cultural signification. If anything, the switch's instantaneous operation and action at a distance dramatized the device paradigm, at least for the generations that marveled at the spread of electricity. To the extent that its diminutive profile nonetheless monumentalizes a cultural preference for concealed production, the light switch, with its disproportion between physical effort and visual impact, is the device paradigm made emphatic (figure 2.15).[55]

DOMESTIC THEATER

Among the greatest gifts that electricity has bestowed on domestic life, is the incandescent electric light.
A. E. Kennelly, 1890[56]

As electrical service spread in the late nineteenth century, switches proliferated. Even private homes began to be stippled with them. Writers in the 1890s evoked "an age of buttons" wherein formerly lumbersome labors became effortless for all. "We are nowadays getting to do things easier and further off than we used to," a newspaper editorial suggested. "There is a button under every finger."[57] Buttons in homes hardly compared to those pressed in public. Nonetheless, they contained a particle of that drama, which helped ease electricity's entry into the home, rehearse gender roles, and make possible a new sense of self in relation to domestic space. In doing so they took on a new cultural role characterized by a movement from the concentrated power under the politician's fingertip to the dispersed power of buttons pressed by countless citizens in their own homes (figure 2.16).

Here, on the surface of a small plate mounted on a wall of the home, the mobility of electricity—the physical trait that distinguished it from any prior form of energy—was held in suspension, awaiting command, ready, almost eager, to consume itself in turning a motor or lighting a room. A single light touch immediately connected a modest room in a modest house with the cataracts and coal mines, turbines and dynamos, relays, and wires that increasingly powered America

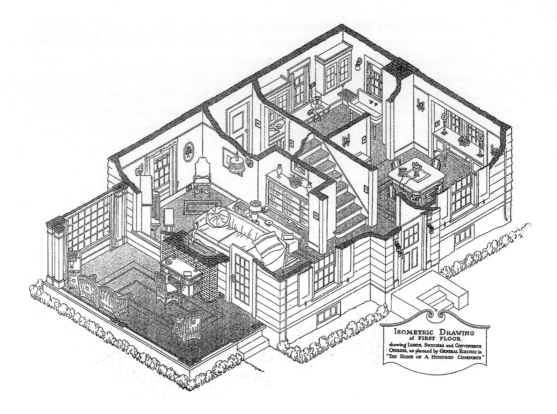

2.16
Makers of lighting equipment, sales representative, builders, and architects worked in the early twentieth century to standardize switch locations and to rationalize domestic activities and movement through the home in relation to electric light.
The Home of a Hundred Comforts (Bridgeport, CT: General Electric Co., Merchandise Department, ca. 1920), n.p.

in the early twentieth century. As a historian of technology put it, electrical energy's "chief significance has been to place power, great or small, in the workman's hands or at his elbow," a claim that could easily be reformulated to include the hands and elbows of homemakers.[58] The switch became a crucial interface between ordinary people and an all but invisible infrastructure, between potent forces of nature harnessed by the latest technologies and the day-to-day doings of everyone. It was the banal object wherein the juggernaut of modernity became the stuff of everyday life.

The space of the switch was most immediate and vivid with electric light, but in the enthusiasm of the day it also extended to other uses. With household chores accomplished by pressing a button, the utility of one's time became more concentrated. Activities that once required much time and labor were newly seen as requiring little more than pressing several buttons. An instant of her attention could thereby substitute for many of the housekeeper's hours formerly spent in actually doing things around the house. Moreover, with a few taps from her finger, the housekeeper could complete chores that just a few years earlier would have required different workspaces within the home. The switch minimized both the time and the space necessary for the completion of domestic activities. This labor-saving notion of button pressing constitutes a functional rather than a visual or territorial conception of the space of the switch.

As houses were outfitted with electrical switches, speculation arose around the turn of the twentieth century regarding the possibility of a button-festooned home, with gates, lights, shoe-cleaning doormats, heated dining tables, serving trolleys, entertainment centers, and more all operated by a touch. In numerous accounts, the home became a place where "the pressing of buttons starts unseen hands to working. By very strange and rapid processes, food is prepared, cooked, served, whisked away and all traces of it removed."[59] Other services, such as decorating advice or private counseling, could summon a new profession of "social prompters," whose assistance could be activated with a switch.[60] In promising such instantaneous gratification, the switch seemed a harbinger of progress. A 1907 report of an exhibition in Paris of the Villa Feria Electra related an authority on the subject who surmised that a good deal of the modernity of modern architecture consisted of the clever use of electricity and the unprecedented convenience of exercising one's will at the touch of a button.[61]

The push-button "house electrical" was so common a trope that it was soon ripe for parody, heir to earlier caricatures by illustrators such as George Rowlandson or Robert Seymour commenting on the new energy technologies of their own day. Certainly the funniest parody remains Buster Keaton's 1922 short film *The Electric House*. In this film, Keaton is accidentally awarded a diploma in electrical engineering, though he has been studying for a degree in hair dressing. He is immediately hired to electrify a home. While the patron is away, Keaton installs a number of clever devices, including moving stairs, an electric dishwasher, a trolley that serves meals, a bathtub that glides over to a bed, a mechanical hand to reshelve books, and several conveniences for the billiard room. The patron returns home and is duly impressed with his new electrical amenities. But at the same time, the diploma's rightful owner, who was initially denied the job when he inadvertently presented Keaton's diploma, unravels the mystery and sneaks into the house to sabotage Keaton's work. Soon all the switches are operating the wrong equipment at the wrong time, to riotous effect. Every button pressed only makes matters worse until Keaton, in despair, ties a rock around his neck and throws himself into the pool and sinks to the bottom, only to see all the water drain out as the lever that automatically fills the pool is switched—manually—by the patron's daughter to empty it, then switched again by the patron to complete the suicide, and then switched yet again to rescue Keaton, who in the meantime has fallen through the drain and finds himself at the end of an outfall pipe, rejected at the film's end by the urban infrastructure he had pretended to master (figures 2.17 and 2.18).⁶²

2.17
Buster Keaton presses the button to operate the electric stairs, but things will go hilariously wrong as a result. Film still from *The Electric House*, directed by Edgar F. Cline and Buster Keaton (First National Pictures, 1922).

2.18
A modern life of pressing buttons is parodied in Jacques Tati's 1967 *Playtime*. Film still from *Playtime*, directed by Jacques Tati (Specta Films/Jolly Films, 1967).

Keaton's film registered a certain anxiety that arose regarding, on the one hand, an invisible infrastructure that guaranteed the switch's operation, and on the other hand the switch's indifference to who was controlling it. Some early twentieth-century authors made use of this unease, with fictional switches used to suggest that all was not what it seemed, to dramatize a plot twist with a sudden revelation, to express vulnerability, and to serve as a metaphor of the tension between seeing and being seen. In the stories in which electric light figures in particular, characters inch along walls in darkness and, at the press of a button, rooms blaze into light: secrets are revealed, suspicions confirmed, relationships upended, and plots leap forward dramatically, if also a bit formulaically: "The wiring of the house, she had already noticed, with the quickness of an expert, was both thorough and modern. Any moment the turning of a bedside button might flood the room with brilliant light and leave her there, betrayed beyond redemption," runs a representative passage.[63] Previous conventional analogies of the moral or psychological profile of characters to the house they lived in—think of Poe's "Fall of the House of Usher"—were modernized in such stories by equating electric light's visual transformation of a space to a character's awareness of a situation. Thus, to the gothic motif of a brooding soul rooted in a gloomy house was added the transformative moment of a literal enlightenment, the sudden, jolting extension of self that was triggered by the otherwise unnoticed switch. The switch was a literary pivot—bearing on both vision and touch—whereat knowledge dilated either to empower or to disempower a character.

WOMEN AND SWITCHES

**She has been accustomed to turn a switch, lighting the lamp immediately.
She presses a button and a bell rings instantly.
Alice Carroll, 1923**[64]

Women were usually the ones pressing buttons, pulling cords, or flipping switches in the home, often motivated by the promise of diminished household chores (figures 2.19 and 2.20). To many observers it seemed inevitable that "housework of the future will be carried on by the turn of a switch."[65] Dr. Lucy Hall-Brown, a physician active in research on electrotherapeutics, told an 1894 meeting of the Brooklyn Woman's Club that push-button housework "was a new form of the doctrine of the emancipation of woman," a prospect the group greeted enthusiastically. Without electricity one found "poor Bridget, hot and tired, … tugging a heavy pail of coal up a stair." One day a workman wired the house, and

2.19

A woman switches on the light before sitting down to her needlepoint work. Matthew Luckiesh, *Light and Color in Advertising* (New York: Van Nostrand, 1923), plate 1, opp. p. 1.

2.20

A servant switches on the lights in preparation for dinner. Advertisement, Lightolier Co., *Electrical Merchandising* 25, no. 6 (June 1921): 6.

Presto! Change! Bridget and the house have become things of beauty and joys forever. … Bridget's temper and the kitchen have cooled together. She comes down stairs in the morning, touches a button, and the coffee is steaming hot; another button, and the eggs are beaten, and still another, and the meat is chopped.[66]

Despite the presumption that efficient domestic labor yielded beauty and joy, and however much they eased labor, push buttons were not likely to bring political emancipation for Bridget, who, along with the "Hilda or Katie" also mentioned by Hall-Brown, personified recent Irish and German immigrants running working-class households or finding work as domestic servants.[67] But the tidy formula made clear the extent to which the expansion of consumer society required not just the use of new commercial products but also naturalized a set of social relations with products, lubricated by routine and intimate encounters with electrical technologies. Not only were Bridget's domestic chores satisfied by nothing more than touching a button, her emotional temperature and the kitchen's air temperature ran along the same circuit, both controlled by the same button.[68]

In addition to pressing buttons, women—female servants in particular—might also be replaced by them. Switches were a new option for bachelors unable to afford a housekeeper or whose current housekeeper, having stayed out late the evening before, was unable to complete her daily duties. A 1908 account of "the magnificent victory of mechanism over maid" featured such a man, a Mr. McMurtry, who had his city home electrified and was able to get on without a servant, enjoying comfort and convenience by electrical means: air warmed, bath drawn, breakfast ready, all, of course, "accomplished each morning merely by your pressing a button." The switch was the emblem of his time, McMurtry mused: "It is the age of the push button." All that was left for human labor, it seemed, was reproduction: "But there is no button we can press to make the living," he concluded on an inexplicably disappointed note.[69]

Even when they did use them, women were portrayed as being confused by switches. An 1895 story described Frank, an electrician, who installed numerous circuits in his home for as many uses as he could think of. He carefully numbered all the switches so that Mary, his wife, would know which switch performed which function. She could never keep them straight, however. After pressing a button to lullaby her baby to sleep, Mary tried to switch on her electric foot warmer. Suddenly the baby was squealing again, awakened by a noise that rang out from downstairs. "You got the wrong switch. That's the burglar scare in the front hall," Frank chided her. "Oh, dear, what a thing to marry an electrician," Mary sighed, "I shall never remember all these switches."[70] Although no house at the time was so abundantly speckled with switches, the caricature demonstrated a cultural familiarity with switches and the manner in which otherwise neutral technologies may be received in gendered terms.

The problem that women allegedly faced was that the operation of switches was so effortless and their benefit so commodious that they were lulled into an unthinking complacency. In a meeting of the Ohio Electric Light Association, a company agent recalled how his daughter told him that the light switches in their home had stopped working. After some inquiry, he learned that she and her mother had forgotten to switch off the iron once they had finished laundry chores. "I went there and looked at the switch, and exclaimed, 'My dear girl, why didn't you press the button?' They had not pressed the button to turn off the current," he reported, to the amusement of his male colleagues. Although the subject of discussion concerned damage to an iron's heating coils when operated for an extended time, the agent took a moment to express his bafflement at women's frequent failure to perform even this trifling act of managing electrical tools.[71] In this and numerous other instances, the popular metaphor of electricity's magic—an agreeable ignorance of how it worked—was employed to reinforce long-standing attitudes toward female technological incompetence.[72]

Utility representatives speculated that women's weak grasp of electricity was hindering its spread and the growth of the industry. Women understood gas lighting, some said, because they engaged with it more closely, trimming wicks, cleaning chimneys, kindling burners, and adjusting valves, and, of course, they could see the flame. No mystery there. But with electricity "they but turn on a switch and heat comes from an unseen and unknown somewhere. They mistrust it." Although women relied more and more on electrical controls, the "ease and celerity" of operating switches, according to the usually male authors, fostered a sense of estrangement from their own homes and, consequently, a degree of apprehension regarding electricity.[73] As a result, the professionals in the electrical lighting industry, who were overwhelmingly male, saw women as both a weak link in sales and a target demographic for corporate growth. To play their part in the continued expansion of the industry, women would need to learn more about what exactly the switch does.[74]

At the same time, some observed, what was termed the "push-button habit" was also leading women to want even more chores done with even greater ease, to the extent that the considerable progress in electrical control equipment was felt to be falling short of skyrocketing consumer expectations.[75] In response, and to guarantee further growth in domestic electrical service, the National Electric Light Association (NELA), a consortium of trade interests, established in 1921 a public relations subcommittee, the Women's Public Information Committee. Women had already been playing an active role in purchasing decisions, and retail shops had refashioned themselves to appeal directly to female customers. NELA saw that trained home economists, nearly all of whom were women, could

2.21

The Habirshaw Company, realizing that homeowners knew little about wiring, aimed to expand its business by making its name familiar to consumers, an approach taken by a number of other manufacturers of otherwise unseen building products. In an advertisement that ran in a popular periodical, the company explained what lay behind "the magic of switch." Advertisement, Habirshaw Co., *Saturday Evening Post* 193, no. 41 (Apr. 9, 1921): 96.

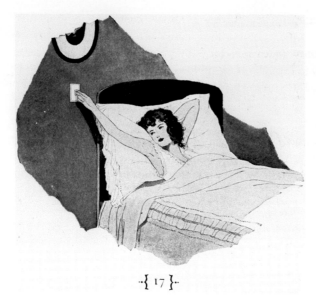

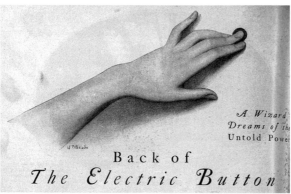

2.22

"Of all the switches in the house this is the most important if you are awakened by any noise or alarm in the night. ... A single move of your arm and flooding light makes you master of the situation." The ability to switch the light on or off while still in bed is part of the convenience in *The Home of a Hundred Comforts* (Bridgeport, CT: General Electric Co., Merchandise Department, ca. 1920), 17.

2.23

Utilities hoped that explaining to women what lay "back of the button" would make them amenable to occasional rate hikes. The implication was that electrical amenities were too easily gained and, as a consequence, disregarded. Charles Steinmetz, "Back of the Electric Button," *Good Housekeeping* 76, no. 5 (May 1923): 48.

become emissaries of electricity and further a female clientele.[76] Noting that women were the ones to complain about billing increases or service interruptions, the committee advised utility companies on better ways to communicate with female customers and how to explain the complexity and strict safety measures of electrical service, which would in turn justify higher prices. The companies decided to educate women about "what lies behind the button." As Miss Bursiel, "chairman" of the committee, put it: "Many of your women customers don't even know where your power plant is located and whether it is an ice plant or an electric power plant. About all they know is that they press a button and the light comes on." Learning the history of inventions that underpinned the industry, the time and investment that went into building power plants, and the whole infrastructure of electrical generation and distribution, Bursiel continued, would lead women to agree to pay more for their electric lights.[77] To aid this effort, NELA commissioned the Rothacker Film Manufacturing Company to produce the promotional film *Back of the Button*, featuring "Mr. Kilo Watt," that explained in simple terms the production and distribution of electrical power and aimed ultimately to educate female consumers about electrical service (figures 2.21, 2.22, and 2.23).

THE TECHNOLOGICAL MUNDANE

The properly citified citizen has become a broker dealing,
chiefly, in human frailties or the ideas and inventions
of others: a puller of levers, a presser of the buttons of a
vicarious power, his by way of machine craft.
A parasite of the spirit is here, a whirling dervish in
a whirling vortex.
Frank Lloyd Wright, 1932[78]

Focusing on the switch highlights a novel but increasingly important way in which men and women relate to their immediate environment, one that goes well beyond the idea that architecture simply contains or accommodates functions. People perform within space—in terms of both carrying out social roles and satisfying functions—but increasingly in the early twentieth century they interacted with space, with the technological systems modernity was weaving into their surroundings. Interactions are necessarily accomplished by controlling flows, whether of energy, temperature, water, other people, or information. After doors and windows, the light switch represents a major step forward along the path to fully interactive environments. The switch allowed space—visual space, at least—to be as readily mustered as any other commercial good. In making any particular space more amenable to a wider range of purposes and for extended hours of operation, the switch also helped make spaces less particular in regard to each other. A well-lit workspace needed no longer to cleave to the windowed edge of a building. Easily accessed electric light made space more fungible. Along with the electrical lighting apparatus it was part of, the switch was an engine of spatial commodification.

As the electrically illuminated room gained functional autonomy from natural cycles of night and day, it became a reservoir of perceptual availability, in the sense described by Martin Heidegger, which also served as a starting point for Borgmann's notion of the device paradigm. Heidegger asserted that the technological transformation of a naturally occurring object or process created a "standing reserve" or supply of potentiality—trees conceived as timber, for example. In turn, these manmade conditions become a second nature and for all intents and purposes appear to have always been so, even though they are demonstrably artifacts of history.[79] Their artificial character dissolves behind a veil of conceptual convention and routine practice; the world, in turn, comes to be understood only as a set of resources to fulfill human needs. To the extent, then, that the switch both transfigured a room technologically and at the same time masked that change, it helped introduce a new, modern space characterized by its responsiveness to an individual's will to see. By mechanizing perceptual potential in this way, the switch became a means by which a space was made modern.

Rather than possessing fixed formal traits, the modern space of the switch was distinguished by several factors: its instantaneity and its amalgam of ignorance and agency. In terms of ignorance, the switch, in keeping with the device paradigm, introduced a gap between the unprecedented convenience of easily controlling anything, from a lamp in one's home to acres of fairgrounds, and an understanding of the means for doing so. Massive machines and marvelous effects were willed into action without the operator's having the least idea of how these things were accomplished. Indeed, ignorance of the mechanism was exactly what underpinned the wonder so many experienced when switches were thrown. The proliferation of electrical devices in the decades around the turn of the twentieth century led to an environment surfaced with switches and controls behind which loomed a realm of otherwise invisible machines. Confidence regarding the switch's operation allowed individuals to accept the continued permeation of technological systems into the built environment and agree to the ensuing asymmetry between operating them and understanding them. The switch made unknowing routine.

On the other hand, the switch returned a sense of agency to its users. It brought the workings of an otherwise hidden infrastructure pulsing with mysterious and potentially deadly energy back into the jurisdiction of human volition, making fingers a crucial element of electric lighting's symbolic system. It installed a human figure into what the writer and historian Henry Adams described as the icon of the era, the electrical dynamo, which he first saw at the 1900 Exposition

Universelle in Paris. Although man-made, it seemed to Adams "a symbol of infinity," comparable to the divine power of the Virgin Mary. Adams focused on the dynamo as an "occult mechanism," a means of producing a force incommensurate with other forms of energy and taxing the mind's ability to comprehend it. He likened the dynamo to faith: both operated by absolute fiat rather than reasoned dialogue, which in either case led to "a sort of Paradise of ignorance" regarding the actual means by which an action was executed. Along with other developments such as X-ray photography, humankind had entered "a new universe which had no common scale of measurement with the old," Adams wrote. Had he even considered it, Adams might well have dismissed the switch as a trivial element in the light of infinity. However, in the eyes of the public, fixed as they were on politicians on podiums pointing fingers, the switch was precisely the hinge between that infinity and mortal being, between an occult mechanism and a handy convenience. It returned a sense of scale to the infinity that Adams identified, a place for humans to reassert their will in a technological universe otherwise indifferent to their existence in it.[80]

In terms of instantaneity, the switch stood in contrast to other fractions of a second. In his 1994 study of depictions of flashes of emotional intensity, the eminent German literary scholar and essayist Karl Heinz Bohrer discerned two types of such moments. One was an epiphany, a sudden and significant insight. The epiphany stood outside historical narrative and gestured toward a verity or transcendent truth; it was a common feature of German idealism. The other moment he termed a "sudden instant," which arose with early twentieth-century writers such as James Joyce, Virginia Woolf, and Marcel Proust. This modernist "moment without duration" insinuated nothing ideal and was either indifferent to or ignorant of metaphysics altogether. Without a transcendent referent, the value of the modernist instant was "the happiness of perception itself," the sheer pleasure of an uninstrumentalized fascination.[81] As another, material form of suddenness, the light switch shared traits of both instances of literary instants. On the one hand, it was a sensory pleasure on its own terms, like the "sudden instant" of modernist work: the instantaneous transition from darkness to light was precisely what attracted crowds to button-pressing ceremonies. On the other hand, the shift between the binary visual states of darkness and light engaged a historical narrative regarding an ideal of progress. It was therefore like the "eternal instant," though not in terms of pointing to a preexisting transcendent truth. Rather, it actively helped construct the modern condition itself *as* transcendent and progress as inevitable (figure 2.24). Unlike these other instants, however, the effect of the light switch could be and often was, in the case of expositions, seen night after night and, in other settings, experienced at will. Through this repetition, the light switch

2.24
A nearly palpable sense of imminent change is evident in this portrait of an instant. In this General Electric advertising photograph a woman is shown pulling the chain on an electric lamp, ca. 1908. Library of Congress Online Catalog, Prints and Photographs Division (https://www.loc.gov/item/2011660037).

installed an experience of suddenness within the fabric of the everyday. It gave a visual correlate of the epiphany—seeing the light—a material foundation and an unprecedented ubiquity. Rehearsing these novel conditions, turning on the lights every day as needed, was the primary way to make such uniquely modern feelings familiar and, eventually, natural.

As the novelty wore off, the switch's annihilation of space—instantly and at a distance—faded fast to unthinking habit. Electric light switches soon became banal and the enchantment of instantaneous visual transformation became routine. The gauze of habit that settled over switches knit their former fascination into the everyday fabric of modern lived space. We can thus speak of a "technological mundane" in contrast to the "technological sublime," a pleasing sense of stupefaction felt when one is faced with technological wonders such as vast industrial plants or enormous turbines, unprecedented speeds, immense machines, or titanic structures that initially strike one viscerally and precipitate swiftly into aesthetic insight.[82] The technological mundane is the residue of the technological sublime, something that was a wonder to a prior generation or even to oneself in the course of an initial encounter but that comes to be a commonplace. It signifies not a mere banality but the contraction of awe into amusement. The comic misadventures of button pressing described earlier in this chapter were artifacts of the switch's diminishing significance. Even the golden keys and hoary pronouncements appear to be strained efforts to compensate for a condition that was already growing stale by the start of the twentieth century. In this sense, the switch miniaturized the sublime. With its nonchalant operation, its small size, its housebound scope, and even its dollop of agency, the switch returned the sublime as a tiny, replicable spectacle. Even today, and despite the switch's complete naturalization, the expression "at the flip of a switch" retains a trace of excitement at the possibility of instantaneous transfiguration.

However prosaic, the technological mundane is far from inert. Its ordinariness is central to its cultural power and to its contribution to the understanding of modern space. No further proof is required of this transformation than the sense of powerlessness, sometimes bordering on panic, when we need to light a room but are unable to find the switch. The miniscule transition from dark to light becomes unbridgeable. The feeling can be disquieting. In describing situations that give rise to unsettling experiences, Sigmund Freud mentions the case of moving through a darkened room, searching for the light switch. Repeatedly, almost obsessively, one retraces the same fruitless route, analogous to returning again and again to the same spot along a misty mountain path or circling round and round the same streets while lost in a strange city, all experiences leading to "the same feeling of helplessness and of uncanniness."[83] The uncanny feeling Freud describes arises not as a sensation separate from the familiar but because it displaces something familiar with something alien. The historical formation of the switch as an ordinary aspect of everyday life induced that sense of familiarity with the benefits of new electrical technologies and a confidence in their continued reliability. When the switch is missing, space itself can feel unwelcoming.

As new technologies were assimilated, routinized and miniaturized, momentous shifts in the prevailing patterns of everyday life were experienced in such slight and seemingly trivial ways as casually pressing a button to light a room. Whereas Karl Marx's observations regarding the annihilation of space and time was frequently cited in reference to the ruptures in social scale triggered by advances such as the railroad and telegraph, such changes were often experienced in minute and even intimate ways that are all but forgotten today. The switch was one such detail that nevertheless played a major role in projecting a sphere of authority, asserting a new, technically mediated political geography, and generating a new and distinctly modern space founded on the willed transition between darkness and light.

3

DRIVING THROUGH THE AMERICAN NIGHT

Of all the skills demanded by contemporary civilization,
the one of driving an automobile is certainly the
most important to the individual, in the sense at least
that a defect in it is the greatest threat to his life.
James Jerome Gibson and Laurence E. Crooks, 1938[1]

INTRODUCTION

Among its various imperfections, the Ford was one-eyed,
and our little light did not cast its beams very far.
We got tangled up into a long line of camions, with blinding
headlights, quite extinguishing us as we hugged the
right side of the road.
Edith Louise O'Shaughnessy, 1918[2]

The automobile headlamp is a modest device, a practical necessity introduced late in the nineteenth century, swiftly standardized in operation and assimilated to driving codes and conventions and just as quickly forgotten, except when it fails to work. Yet the headlight—the light produced by the headlamp—introduced a luminous space with length and breadth, an inside and an outside, a top and a bottom, and an immediately recognizable form and character. Unprecedented in its ubiquitous mobility, the headlight soon overran other spaces, such as city streets and country roads, or swept late at night across bedroom walls. The space carved out by the headlight was a distinctly social space: designed, debated, feared, regulated, evaluated, adjudicated, and eventually accepted. As with other spaces, it conditioned the behavior of its occupants relative to their specific position within it. Inside the headlight, drivers and passengers found themselves at the vertex of a luminous cavity moving through a mantle of dark. Outside, at the headlight's periphery, other drivers and pedestrians were insulted by the visual violence of sudden glare and the peril of collision. The headlight's social dynamism was grounded in its conflict with these other spaces and in its establishment of new subject positions, premised on acquaintance with new sensory methods. Pedestrians learned to overcome "sudden blindness," the ephemeral disability caused by the glare of oncoming headlights, while drivers learned to master a novel visual world (figure 3.1).

RUSHING IN THE DARKNESS

Along the South Country Road a rumbling sound approaches
rapidly nearer and nearer. What is it? A motor car! In an instant,
with a roar, and with flashing headlights, it is gone.
Sarah Diodati Gardiner, July 28, 1908[3]

Well before electricity became available, headlamps were common features on moving vehicles such as carriages, bicycles, and locomotives. Typically, on carriages, an oil lamp was hung from a mount at the side of the cab and encased in glass to protect the flame. With neither lens nor reflector, the lamp cast a dim circle around itself, with a fraction of the light falling on the road ahead. Reflectors could direct the light, but it was still feeble and diluted over a wide arc. For their part, bicycle lamps were dim, awkward to install, liable to sway greatly, and easily extinguished (figure 3.2).[4] Riding a bicycle at night, particularly along unlighted and unpaved streets, was difficult and motivated only by pressing need or an outsized sense of adventure. Locomotives, in contrast, carried substantial headlamps intended to provide ample warning of a train's approach rather than light up the rails ahead (figure 3.3). Kerosene, used starting in the 1860s, did not produce an especially bright light, but later lamps, such as electric carbon arc lamps from the 1880s or acetylene lamps some twenty years later, were powerful. These lamps substantially reconfigured the view, but their use was limited to rail lines, which were isolated, fenced, or otherwise marked off from routine circulation. Seen from the side, the locomotive headlamp was a spectacle: its "dazzling glare … shooting like a meteor from the Plutonic abyss, is wild and awful," ran one account.[5] Encountered directly, however, it was usually deadly. Flore, a character in Émile Zola's 1890 novel, *La Bête humaine*, whose setting is contiguous to a railway line, is long drawn to the danger of approaching headlamps and finally commits suicide by standing in front of one. A majestic moose, in a story from the American clergyman and writer Henry van Dyke, rears up to smash an oncoming headlight, with similarly unhappy results.[6]

3.1

"What is it?" General Electric products helped drivers decode the new night landscape. Advertisement for Mazda automobile headlamps in the *Stimulator*, October 1919, 5. The *Stimulator* was a promotional and sales periodical given to distributors by the National Electric Lamp Association, NELA Park, Cleveland, Ohio.

3.2
Illustrators found useful pictorial precedents for representing the headlamp's projected cone of light in the work of such artists as Aubrey Beardsley and others influenced by Japanese woodcut prints. Advertisement for Tally-Ho Bicycle Lantern, *Cosmopolitan* 20, no. 6 (April 1896).

3.3
Seeing the track ahead beneath the locomotive's powerful projected beam of light. Captioned "The electric headlight in operation," in "A Successful Electrical Headlight," *Electrical World* 16, no. 18 (Sept. 27, 1890): 221.

Obtaining the point of view of the locomotive engineer in the front cab was rare, but by the late nineteenth century writers on a variety of subjects were seeking ways to characterize the faster tempo of modern life, often finding it in new forms of transportation. In such accounts, speed was especially vivid at night, when darkness shrouded distance and diminished motion parallax, the apparent differential movement of fixed objects against a background that provides perceptual cues to depth. With the spread of gas and electric lamps, speed was rendered an incandescent spectacle, a phantasmagoria of moving lights. To the mind it was a puzzle of trembling flicker, glare, and glint dissociated from and even dissembling the substance and identity of the otherwise coherent objects on which it played (figure 3.4).

One of the best descriptions of this visual voyage appeared in *Scientific American* in 1902. The author, an editor of the journal, began with an account of the demands pressing on the engineer's attention. As the engineer's guest, however, he was at leisure to take in the "pure sensationalism" of the experience and its appeal to his entire body. The train roars like a "strident orchestra"; the "great mass of the engine vibrates and lurches and rolls" and threatens "to rend itself into a thousand fragments." For the eye, the effect was otherworldly: the headlight, aligned with the driver's vision, was focused, bright, and horizontal. Everyday objects caught in it were reconfigured in terms of the constituent reflectivity of their component surfaces, while independent light sources hurtled toward him. All perception of distance was extinguished and the ability to identify objects underlying the glimmer was greatly curtailed; they could be recognized only at the last instant, if at all:

> By day, objects approach slowly out of a far perspective; but by night they rush at you out of the near darkness in one mad whirl of ghostly shapes, punctuated by horizontal, rocket-like streaks of fire—the signals and station lights.[7]

At high speeds reflections emerging from "the near darkness"—vision evacuated of its contextualizing daytime perspective—acquired their own rhythm and luminous sequence, perpetually flowering into lucent form. The most visually intense moments were experienced when one passed through stations, where various lamps were fused by the observer's speed into a new dimension of light:

> You sweep down upon a mass of white lights, red lights, headlights, whirling hand lamps, dwarf signal lights below, and arc lights above, with two or three switching locomotives to heighten the crowded effect! ... at the very front end of this roaring cataract of steel and fire.[8]

Although few would witness the world at such speed, the mass production of automobiles and the steady improvement in automobile lighting equipment meant that thousands and soon millions of drivers could venture out in their cars after sundown to occupy a similar position propelling a cone of light into the night.

3.4
The locomotive engineer's view on entering the rail yard. Here, switch lights, headlights, and other lamps have been hooded in accordance with blackout regulations. The thick streaks of light are from locomotive headlights; the thinner streaks are from switchmen's lamps.
Activity in the Santa Fe R.R. Yard, Los Angeles, Calif., Jack Delano, photographer, March 1943. Farm Security Administration/Office of War Information Collection, Library of Congress Online Catalog, Prints and Photographs Division.

Modeled on earlier forms of transportation, the first automobiles carried lights of some type. Often they were hung from the side of the car so that the driver could readily adjust them. Headlamps, either one or two, were mounted on the front of the vehicle and fueled by oil or acetylene, which flamed brightly and was resistant to being extinguished by wind. Electric headlamps were introduced at the end of the nineteenth century, but they were more delicate than and, until about 1910, not as bright as acetylene lamps, which remained a viable option through that time. As filaments became brighter and more durable, and as electrical systems with generators of ever higher voltages became more sophisticated, reliable, and standard a feature across manufacturers, electric headlamps became the norm, even though they were often considered add-ons to the vehicle's base purchase price.[9] Many drivers put together a composite system with, for instance, electric headlamps supplied by the manufacturer and a custom, side-mounted acetylene "searchlight" that could be aimed wherever the driver chose and removed to illuminate flat tires or to make other unplanned repairs.

The number of registered automobiles skyrocketed, from some 8,000 nationwide in 1900 to over 450,000 ten years later. With Ford's affordable Model T in production since 1908 and the first federally funded road legislation in 1916, their numbers soared to more than eight million by 1920.[10] The flood of cars and their careless drivers brought with it "a number of evils," including smoke, unpleasant odors, dust, the "needless tooting of the horn," and "flaring acetylene headlights," which were "an actual annoyance and possible source of danger to pedestrians and horse-drawn vehicles." Headlights from dockside cars were even distracting to ferry pilots.[11] As early as 1900 the automobile was labeled "the new juggernaut" for its speed, mass, freedom of movement, and seeming indifference to darkness.[12] Despite earlier successes of the Good Roads movement, most thoroughfares were unpaved and populated by carriages, animals, and pedestrians, who had little cause for caution when crossing against slow-moving traffic. Regardless, drivers in ever greater numbers followed their headlights into the night.

NIGHT DRIVING

I never knew what real pleasure "night driving" was until I installed the electric lights.
Harry X. Cline, M.D., 1912[13]

The industry initially promoted leisurely rather than utilitarian uses for automobiles and emphasized driving's thrilling appeal to all the senses: the wind pressing the face, whistling and whooshing sounds as the car flew along the road, the accelerated pace of passing scenery, a varied somatic awareness of wheels skittering over different road surfaces, the rumble of the motor, and swaying sensations that intensified at turns. To these, night driving added stunning visual traits (figures 3.5 and 3.6). The world wore a new aspect at night, and even familiar scenes were changed and vision refreshed. For all but the most practiced night driver,

> no matter how well he may know the road by daylight, he is often utterly at a loss to recognize familiar objects and will even pass the place to which he desires to go without being aware of it, when traveling over the same road after dark. Objects by the roadside have an unnatural appearance and seem out of proportion; what appears as a dark patch in the road may be either a pool of water or a depression, and light colored objects by the side of the road may even be taken for the road itself.[14]

3.5
"A beautiful stage effect observed in night driving." The effect is composed of picturesque indicators of irregularity, distance, and motion, all heightened by driving at night. Joseph Tracy, "The Pleasures and Perils of Night Driving," *Harper's Weekly* 52, no. 2666 (Jan. 25, 1908): 27.

3.6 (following page)
Even in more formal, urban settings, the headlight generates its own picturesque effects, including a changing sequence of scenes. "Beauty in Night Motoring," advertisement, Prest-o-lite Co., *Good Lighting and the Illuminating Engineer* 7, no. 10 (December 1912): 520.

Although "unnatural," such effects were deemed different from, rather than inferior to, the same scene viewed by daylight.[15]

The power of night vision was discovered in the unique space generated by headlights: a roughly circular scene filled with objects constantly approaching the car but never coming into contact with it, as long as one's luck held. The scene was structured by certain constants, too, including the circumference of the headlight pushing aside a mantle of dark, a road surface that bisected the scene and always brightened as it approached, a number of minor lights emanating from the dashboard and sidelights, reflections off the windshield or polished parts of the car, reflective materials or actual lamps along the road, and, most menacing, intermittent glaring blasts from the headlights of other cars. Although drivers had always to remain alert to the dangers of driving, exacerbated by the unfamiliar nighttime visual field, the visual and somatic sensations of speeding through the dark combined to engender perhaps the most enthralling entertainment of the time.

Harold Whiting Slauson, an automobile enthusiast and author of many articles and books on cars, noted the growing popularity of driving in the dark: "Night driving seems to be becoming more and more a popular feature of automobiling, to judge by the number of headlight and searchlight accessories to be found."[16] He went on to articulate "the joys of night driving." The "pleasure of motoring," wrote Slauson, "is doubled when even familiar scenes are traversed after sundown." The reason for the doubled pleasure was itself twofold: headlights and speed. Headlights rendered forms in a novel fashion even as daytime habits made them all but invisible. The strong sidewise light exaggerated forms and cast long shadows that melted into the surrounding darkness. Speed simultaneously foreshortened objects and condensed observational time: objects appeared vague at first, perhaps just a flash from a polished surface, bloomed gradually into nocturnal form, then sharpened for an instant and, just as swiftly, were gone, to be quickly replaced by another baffling flash from the darkness ahead. Even trivial roadside elements, said Slauson, "assume an entirely new aspect when touched by the soft moonlight or the flickering, flaring headlights that bring approaching objects into a relief all the more strong because of the contrast with the shadows into which they quickly fade."[17]

In turn, peering into the headlight sparked a new kind of attention that was simultaneously vigilant and relaxing (figure 3.7). Unremitting focus was requisite since even minor lapses in attention could have devastating results. Indeed, insufficient attention was common grounds for "contributory negligence" in legal cases concerning vehicular accidents.[18] As described in 1905 in an English periodical, drivers operated in a state of constant emergency: road surfaces changed from one instant to the next; traffic flowed with, against, and at all angles to their own direction of travel; they had to be alert for curious children, sleeping cattle, deer hypnotized by headlights, stumbling drunkards, and "even for the lovers strolling along and neglecting the road and its occupants, because, forsooth, they tread on air!" Even passengers ought to refrain from "unnecessary talking" that might distract a driver. Everything on the road had to be quickly identified and assessed for any threat to forward motion. But this purposeful, prolonged gazing into the space of the headlight was also quite pleasant. The visual behavior produced by speed—"when time is much too valuable to spend in thinking"—lifted the driver toward an aesthetic frame of mind: "It is without any doubt this necessity for concentration, this ever-present call for watchfulness, that makes up a great part of the charm of driving a motor," and, along with a mechanical awareness of the car itself, contributed to "the complement of joy to the complete motorist."[19]

3.7
Night driving encouraged an alert yet relaxed frame of mind, the ideal mental attitude for safety and visual pleasure. Film still, *Knights on the Highway* (Chevrolet, General Motors Sales Corp., Jam Handy Organization, 1938).

3.8
In the new geography of electric light, the headlight brings a glint of urban character to the countryside. *Light* 8, no. 6 (Summer 1930), cover. Reproduced with permission from the Archives and Research Center, Museum of Innovation and Science, Schenectady, New York.

Steadily peering into a cone of light, drivers found night driving taxing but also restorative. Repeated scans of sequences of reflected light displaced the day's distracting, random juxtapositions with a singular focus that, despite the necessary vigilance, led to a tranquil frame of mind. As one analysis put it, at night "the rapid passage of the car gives rise to a sense of buoyancy and the reaction therefrom causes a feeling of drowsiness which makes the autoist sleep like the proverbial top."[20] Rather than a result of the driver's purposeful filtering of perceptual stimuli, this relaxing effect was due to the headlight's luminous constitution. Shining unfalteringly forward, it relegated irrelevant information to darkness and thereby eased the chore of concentration: "The inability to see the minute details of grade, surface, etc., leaves the motorist comparatively free from nervous strain," ran a representative view.[21] The headlight, in other words, created by its sweeping exclusions an enduringly observant present that gave rise to a mobile version of what, in another context, art historian Jonathan Crary describes as a "suspended temporality, a hovering out of time" that is the product of rapt attention.[22]

Automobile headlights also helped give a figure to a new geography of night then being forged by the uneven spread of electrification. Whereas cities sharpened their cosmopolitan profile with building, street, and festival lighting, the countryside was recast as an undifferentiated domain of empty darkness waiting to be sliced apart by motorized thrill seekers (figure 3.8). Whereas cities were crowded, country roads were relatively traffic-free, inviting drivers to go faster. The city dissolved the volatile space of the headlight in its wash of street lamps and illuminated shop signs. "The big headlights are for unlighted country roads—not for boulevards. They are not allowed in New York. Why are they permitted here?," asked one Chicagoan.[23] Country roads, in contrast, were dark. Cities were also too cramped to allow the mental effects of night driving to fully emerge, which only unfolded over time as the eyes steeped in the headlight. Moreover, city drivers were goal-oriented. Presuming that aesthetic reception flourishes only in the absence of an immediate objective, Slauson wrote that city driving was a means rather than an end in itself, "a short run to the opera," for instance, that eclipsed the journey with the destination. The overall result was that "this man," the city driver, gathered only "half of the value and enjoyment that he otherwise could obtain from his car."[24] Fewer lights in the countryside thus meant less competition for the driver's attention. Night driving became a palliative for urban seeing, especially suited to "nervous persons or those whose occupation keeps them indoors for the greater part of the day."[25] Once headlights wedged open the night and made nocturnal motoring feasible, drivers went to the countryside to rediscover the darkness that urban lighting had stolen (figure 3.9).

During the day, the car might be hindered by a horse-drawn wagon or by cattle being driven down the road, in an emblematic encounter of work and leisure rewritten in terms of transportation progress. The standard trope featured the driver leaning on the horn, able to leave behind the city but not his impatience. Night driving upturned the equation: the city dweller's leisure was unimpeded but the farmer's rest was broken by the roar of a motor and the sudden sweep of a headlight inside rural chambers. City drivers went to the countryside for "midnight 'tears,'" so named because the speeding cars "tear up space."[26] Nighttime drivers tearing up country roads was a geographic emblem of the class tensions automobiles brought more generally, as Woodrow Wilson pointed out in 1906: "To the countryman they are a picture of the arrogance of wealth, with all the independence and carelessness."[27] Thus, night-blanketed country roads read as the inverse profile of accelerating and diversifying urban life at a time of tremendous infrastructural growth and soaring immigration. In this context, distinctions between city and country night driving became ideologically freighted.

3.9
The dream of unfettered roadways for driving, whether by day or night, taken to its utmost. Lyonel Feininger, *Rekordfahrers Traum oder: Kilometerfraß in den Ringgebirgen des Mondes*. *Lustige Blätter* 19 no. 24 (1904): 20. My thanks to Sherwin Simmons for this image.

THE DANGEROUS SPACE

Euclidean and perspectivist space have disappeared as systems of reference.
Henri Lefebvre, 1974[28]

The headlight also helped the general public become familiar with the idea of "braking distance," the distance a car travels when coming to a full stop, measured from the moment the brake shoe engages the brake drum. Recognition of braking distance remains a distinct, dynamic, and consequential instance of what various writers have described as a more general shift around the turn of the twentieth century, the emergence of a modern sense of space.[29] In that broader context, braking distance might well be the first widely encountered, lived experience of modern space-time, akin to far grander developments but occurring in the everyday rather than in scientific theories or on artists' easels. Passengers might reflect on the novel interchangeability of space and time of train travel but the automobile driver was an active agent in determining braking distance, a dimension that had to be recognized immediately in the presence of obstacles to forward motion. As with other aspects of electric light such as the switch, braking distance entailed a new relationship to space that first had to be recognized, then mastered, and, once sufficiently familiar, internalized as part of life in the modern world (figure 3.10).

Braking distance had already been an important issue for the railroad as stronger engines propelled trains more rapidly and service grew to accommodate greater numbers of travelers going to ever more destinations along an expanding network.

3.10
"Measuring the distance." Coming to understand braking distance in absolute terms of measurement as well as in terms of perceptual experience. Photograph, Bureau of Standards. H. W. Slauson, "Saving the Motorist from Himself," *Safety* 11, no. 5 (November–December 1925): 147.

Train schedules had to be tightened, while hazards, such as grade crossings, multiplied. As a result, "stopping distance," as it was more generally known on rail lines, became a precarious matter. "Every train bears before it a dangerous space equal in length to the braking distance at that particular speed." The problem, engineers routinely pointed out, was not in the braking system but in "human fallibility. ... The majority of accidents upon electric railways come from disregard of this elementary idea of the dangerous space."[30] At night, this dangerous space was an even greater menace because of the train engineer's diminished visibility, and designing headlights congruent with the long braking distances of trains was exacting. For this reason, braking distance and headlights were often studied in tandem, with the latter providing a ready figure of the former.

Automobile engineers reviewed railroad sources to better understand braking distance in cars.[31] They readily agreed that the "two absolutely essential safety appliances on a motor car are the brakes and the headlights—hence the enormous importance of the headlight problem."[32] In time, a consensus formed that braking distance ought to be sufficient to allow the driver to recognize objects on the roadway ahead. As with railroads, the difficulty lay with the driver. While braking distance might be objectively measured by taking into consideration aspects of the vehicle (its speed, weight, brake, and tire quality) and the road (surfaces, curvature, obstacles, weather), the driver had first to notice an obstacle, then recognize it as such, and then react by engaging the brake. The duration required could only be estimated on an average basis. Thus, calculating braking distance was "very much complicated by the entry of optical, physiological and psychological elements." In consequence, braking distance, like vision itself, was commonly understood as an amalgam of subjective and objective factors brought together by mechanical means of transportation to create a new kind of space.[33]

Charged with finding precise measures of liability, courts, based on the expert testimony of engineers, explicitly differentiated "reaction time" from "braking distance." While their sum was the total time it took for a car to stop, a driver could be partially responsible for the latter and wholly responsible for the former, barring other distractions such as glare. Moreover, courts were sensitive to the fact that even objective measures were only partly reliable: braking tests were executed under conditions somewhat divergent from the actual conditions of an accident.[34] The onus for determining what came to be called "assured clear distance ahead" was thus laid on the driver in a host of state regulations stipulating, in the legal scholar Arthur Leff's formulation: "The driver shall at all times operate the motor vehicle at a speed which will permit the vehicle to be brought to a halt within the space of the assured clear distance ahead."[35] Drivers thus acquired legal responsibility for understanding their car's capabilities, the prevailing road conditions, and their own ability to respond to hazards. "Do not overdrive your headlights" became a common admonition of night driving at this time (figure 3.11).[36] In this way, drivers came not only to understand the concept of braking distance, they also owned it. That is, drivers were responsible for a volume of potential vehicular movement that was itself in constant motion. Nevertheless, municipalities, free to exercise their own judgment in regulating automobiles, would regularly require demonstrations of braking distance before settling on matters such as town speed limits.[37] The overall result was that even through the 1920s, there was "no agreement … as to what is a safe braking distance or what constitutes a satisfactory braking test." Needed were national safety standards and, ultimately, mandatory driver education classes.[38]

The first sustained phenomenology of braking distance was offered by the perceptual psychologist James Jerome Gibson, in a 1938 essay on the visual field of the automobile driver. In it Gibson developed a concept of the "field of safe travel," which describes a projected obstacle-free zone that precedes a moving automobile like "a sort of tongue protruding forward along the road" (figure 3.12). By day, this field varied considerably in relation to such things as the anticipated movements of pedestrians and other cars. It was not subject to precise measure and changes from instant to instant, he continued, but was nonetheless "present implicitly in every driver's field. How accurately it accords with reality is another question" that bore on mechanical, physiological, psychological, and perceptual factors. Having internalized a sense of braking distance, drivers projected themselves forward to a "potential location" that, ideally, fell short of a perceived obstacle. The difference between the projected potential location and the perceived obstacle constituted a margin of safety. This margin could be compressed by speed to zero or even into negative numbers.[39]

3.11
"At night the road we can safely drive on moves along only within the path of our headlights. So expert night drivers always remember the rule, 'Keep within your headlight beam.'" The headlight beam constitutes a modern lived dimension of time and space. Film still, *Knights on the Highway* (Chevrolet, General Motors Sales Corp., Jam Handy Organization, 1938).

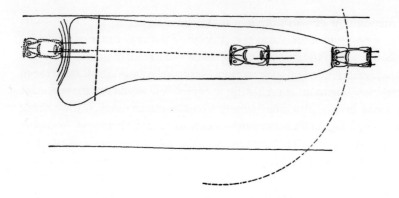

3.12

Braking distance can be defined as "the field of safe travel at relatively high speed." Its contours are in constant flux as a result of changes in speed, visibility, and surface conditions and the presence of other moving objects. At night the visual space of the headlights adds another level of complexity by containing or falling short of braking distance. James Jerome Gibson and Laurence E. Crooks, "A Theoretical Field-Analysis of Automobile-Driving," *American Journal of Psychology* 51, no. 3 (July 1938): 463. Copyright 1938 by the Board of Trustees of the University of Illinois. Used with permission of the University of Illinois Press.

3.13

Braking distance was closely related to speed and visibility. The space of the headlight curtails the driver's view of the outside world, to the hazard of other vehicles, drivers, pedestrians, and horses. "How and Why to Eliminate Glaring Headlights," *Automobile Journal* 64, no. 9 (December 10, 1917): 8.

Gibson mentions headlights only as a restriction on visibility but they can be understood as something more than that. At night, the field of safe travel that is amorphous by day is overlaid by the fixed perimeter of the headlight's cone, punctured episodically by reflections from shiny objects or other lights located well outside the field. With most streets outside cities unlit at this time, obstacles approaching from the side would be recognized only by the harbinger of a moving lamp (figure 3.13). The parallax available to drivers by day would be eroded at night, and strong shadows cast from bright horizontal headlights could prolong the driver's reaction time. Nighttime braking distance, in other words, was, as a qualitatively distinct modern space, not only generated by the forward motion of the car, it was also circumscribed by the limits and special character of the headlights, a product of spinning motor and projected light that was grasped, internalized, and subsequently normalized as a commonplace interval of everyday life.

THE MORAL BLINDNESS OF GLARE

"Glare" is the one word most used in referring to the blinding effect
of high candle-power units. This question probably commands
more widespread interest at the present time than any other problem
within the scope of the illuminating engineer.
Evan Edwards and H. H. Magdsick, 1917[40]

In nearly any discussion of the matter, the speed of the automobile at night depended on the quality of the headlamps, or, to put it another way, speed was directly proportional to the brightness of the luminous space into which the car continually headed. More powerful beams deepened visual space, which translated directly into increased braking distance in space-time, which could be compressed further by pressing on the accelerator, leading to a demand for still more powerful headlamps (figure 3.14). Better headlights and faster driving were reciprocal. Acetylene headlamps were already extremely bright when automobiles numbered only in the thousands, but the glare problem "became especially insistent when electric headlights were adopted for automobile lighting."[41] Typically fixed near the front bumper rather than mounted alongside the driver for easy manipulation, electric lights were aimed directly in the line of vision of alarmed pedestrians and drivers of oncoming vehicles. The combination of more cars, brighter lamps, ineffective shading, inconsiderate drivers and, often, unpaved or otherwise bumpy roads meant that "the annoying and even dangerous glare had multiplied" to an even greater extent than mere wattage might suggest.[42] Standards failed to keep up with the spread of automobiles and, owing to the ambiguous definitions of glare and unreliable means of measuring it, states and municipalities proposed legislation that was "futile and in many instances ridiculous" in trying to manage glare.[43]

Mitchell May, New York's secretary of state, charged with regulating the burgeoning automotive trade, and later state Supreme Court justice, framed it this way in a 1914 newspaper account: "As the speed of motor cars has increased, so have the power and efficiency of the lighting system improved." Headlights had grown so bright that they blanched roads "for several hundred yards, every object being bathed in a flood of brilliant white light."[44] At the same time that brighter lamps made faster driving safer for the occupants of one car, they made it more dangerous for those at the headlight's periphery:

> There are few situations more paralyzing to the traveler of any description than meeting a motor car with powerful headlights. All that is visible to him is a pair of eye-scorching white disks set in the midst of impenetrable darkness. Anything that may be on either side of or behind those lights—man or beast—is absolutely invisible to him.[45]

May noted that dimmer lights were also problematic since drivers would have less scope in their view and thus less time to resolve oncoming objects. To lighting engineers, headlights were an explicit dilemma: higher-intensity lighting equipment has, "in solving the problem of lighting the road, introduced a new and serious problem in that they temporarily blind the driver or pedestrian who happens to come within their angle of action" (figures 3.15, 3.16, and 3.17).[46] Stronger headlights, in other words, led to a set of visual disorders that brought new challenges and required new accommodations to an increasingly routine act of seeing.

3.14
Powerful headlights can give drivers a false sense of confidence. "The driver who was injured was entirely to blame because the blinding glare of his lights caused the trouble. Don't think because you have powerful headlights ... that you'll escape when some accident happens." Night drivers were advised to keep their headlights within legal limits or face the consequences. Headlamp advertisement, Conafore Glass, Corning Glass Works, *Collier's* 63, no. 16 (Apr. 19, 1919): 38.

3.15
The lower half of the body: the ideal occupant of the space of the headlight. A cone of light produced by Roffy lamp design, said to have a "glareless beam." "Glare-Preventing Devices for Headlights," *Transactions of the Society of Automobile Engineers* 9 (1914): 292.

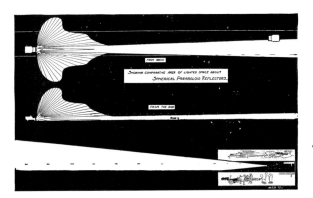

3.16
Engineering the space of the headlight, in plan and section. "Glare-Preventing Devices for Headlights," *Transactions of the Society of Automobile Engineers* 9 (1914): 289.

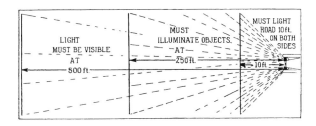

3.17
Analytical plan of the necessary visual performance specifications of space of the headlight. "Light the Road—Let the Heavens Illumine Themselves," *Automobile Journal* 44, no. 5 (Oct. 10, 1917): 9.

3.18

Pedestrians at the perilous edge of the headlight fleeing from reckless night drivers. Captioned "The lives of the hunted," *Auto Fun* (New York: Thomas Y. Crowell, 1905), n.p.

3.19

The ghosts of John and Priscilla Alden, frightened by a fire-eyed creature and its wraithlike cargo. Captioned "Ghost of John Alden," in *Auto Fun* (New York: Thomas Y. Crowell, 1905), n.p.

The professional, legal, and medical literature abounds with conditions of "temporary blindness," "blinding glare," "disability glare," and similar diagnoses, followed by attempts to render in words the precise visual conditions that led to them. In one discussion, for instance, headlights "not only act by producing a transient blinding after-image through excess of light, as by the fact that, in their case, the image is usually suddenly thrown on the dark adapted eye, and therefore produces a relatively enormous effect."[47] Writers often used the nominal term "dazzle" in two closely related senses, as a quantity of headlight brightness and as the visual disarray that results from bright headlights. Night driving became a special field of interest to those studying the etiology of ephemeral visual disorders. "Doctors," began an article in the *British Medical Journal*, "are concerned perhaps more than any other class of the community with the matter of the safe lighting of road vehicles" and ought therefore to protest the less than scientific rationales behind recent headlight regulation.[48]

In more popular discussions, however, visual infirmities were often the result of malevolent others, with bright headlights readily anthropomorphized to denote the threatening gaze of an assailant. A car was likened to a menacing beast creeping up on bystanders at night to "burst upon the vision" and, in one account, the "white ghosts and red devils" of the car lights frightened a man "out of at least seven years' growth" (figures 3.18 and 3.19).[49] A *New York Times* editorial suggested that cars recalled "beasts in narratives of apocalyptic visions," such as Daniel's description of a "beast, dreadful and terrible, and strong exceedingly; it had iron teeth," or St. John's similar description of beasts of metal that "left a very disagreeable odor behind them. … If these machines were beasts, why may not the modern road locomotive not be classed as an animal?"[50] The question had a legal basis. The automobile's legitimacy on the road was analogized to wagons drawn by horses or oxen teams that already occupied public roadways legally. Early court decisions explicitly weighed whether automobiles were, in effect, animals and should be judged accordingly. Although widely acknowledged as "a dangerous instrumentality," cars were ultimately "not classed with ferocious animals." One collection of cases cites a Georgia appeals court case, *Lewis v. Amorous*, on the matter: "It is not the ferocity of automobiles that is to be feared, but the ferocity of those who drive them. They are not to be classed with bad dogs, vicious bulls, and evil disposed mules and the like."[51]

Drivers, in other words, frequently behaved like ferocious animals once they got behind the wheel. Intoxicated—a frequent adjective—with speed, they metamorphosed into "speed maniacs" or "fiends" or "road hogs."[52] Following a headlight into the night seemed to make matters worse. Men ordinary by daylight became grave hazards in the dark. Newspapers carried countless reports of crazed drivers confusing, frightening, and striking pedestrians and animals.[53] The rush of sensations reaped by speeding at night was a solvent that dissolved good judgment and loosened morals. If already so inclined, drivers exploited night driving to further their misdeeds. Where formerly they fled on foot, criminals were newly empowered to speed away in a private car, not only faster but further distanced from observation by the dazzling veil thrown over the car by the street lamps it passed.[54]

For law-abiding citizens too the streaming space of the headlight led to drunken behaviors ranging from reckless to libidinous to illegal. A physician in Chicago identified "the automobile eye" as a state of confusion caused by the unnatural strain of seeing at high speed. Frequent drivers such as chauffeurs were especially vulnerable and "chronically misjudged speed and distances and thus went blindly into collisions. The flickering of the passing road and landscape had much to do with the eye disease." Yet, as a French scientist explained, such drivers continued motoring, seduced into prolonged and unnatural use of their eyes by the exhilaration of driving.[55] The overall result was an "urban madness" of "Wild Cars … and Their Wilder Occupants … the new terror of city life":

> The new terror that haunts the city's ways, lurching around a corner to run down the heedless pedestrian, whirling out of the fastnesses of night to a shower of light, a purr of engines, and a wake of fumes; scudding into the lamp glare with a snatch of ribald song and back again into the black from which it emerged with its gay burden; crashing into posts.[56]

3.20
The headlight lifts the age-old veil over nocturnal behaviors. Captioned "A new light on an old subject," in *Auto Fun* (New York: Thomas Y. Crowell, 1905), n.p.

"Joyriding" was the name given to such behavior, and it was exacerbated at night. Rambling in a realm beyond decent manners, joyriding was night driving gone bad. Women were seemingly unable to resist "the romance of the night riding auto" and submitted, "clandestinely" and against their clearer daytime judgment, to the car's "nocturnal wanderings."[57] The narrative structure of most joyride cautionary tales involved, first, "some half-drunken ruffians," then a woman or two who "foolishly assented" to join them, and finally an accident resulting in injury or death, often accompanied by last words of regret and a lesson for readers on self-restraint (figure 3.20).

With women present, the male driver's behavior only worsened as he tried to outdo himself in recklessness. As the *New York Tribune* reported, "It has become 'the regular thing' for a certain class of automobilists to invite almost any girls whom they may see on the street to ride with them and the fascination of the sport often tempts girls who ought to know better. … 'Joy riders' never run quite so recklessly as when they have thus picked up some foolish or vicious women as their companions."[58] Men were seduced by the power they controlled and "succumbed to the temptation of speed."[59] Consonant with ancient tropes, temptation, in turn, was women's work. Cars were generally gendered female, the purring engine a siren's song. In springtime, for instance, while the world, or at least in this case Long Island, blossomed with thoughts of love, "not a few who ordinarily are to be classed as safe and sane drivers, yield to the temptation of 'stepping on her.'"[60] In this way, mastery of a new technology, the car, a new activity, night driving, and a new space, the penetrating cone of the headlight, was a potent demonstration of male power over nature and over women, if not always over oneself. In much of the literature, women were erotically drawn to the penetrating glare of the headlight as a beacon of virility and motorized male power, a kind of illuminated phallus that led women to swoon where they stood.

When women weren't swooning, they were paralyzed by the headlight's hypnotic grip:

> Ladies on bicycles, when fixed by the brilliant gleaming eye approaching them rapidly out of the dark, become confused and embarrassed, and convey their confusion and embarrassment to their iron steeds, which wobble in consequence from one side of the road to the other, to the imminent danger of the lady riders, and the unspeakable terror of the driver of the car.[61]

Alternatively, women might panic in the headlight, their good judgment apparently melted by the glare, often with deadly results: "After passing a powerful light a rider runs into a Stygian darkness, being temporarily blinded," and thus likely to have or cause an accident, reported the *Motor* in a piece titled "Lamps and Ladies: Dangerous Obstacles." The same article discussed one driver who, dazzled by a

3.21
With a winged wheel in her left hand, at the time an emblem of technological progress, a bare-breasted female figure gestures with a lighted torch held in her right hand toward an automobile with prominent headlights. Poster, Second National Swiss Automobile Exposition, Geneva, 1906.

passing car, gleaned a woman running across the road, but then—"She hesitated, and turned back (when will women give up this reprehensible habit?), and, to avoid her, the driver had to turn the car towards the pavement," striking the woman and throwing the car's passenger to his death. A court found it to be a case of accidental death due to the driver's temporary blindness.[62] Similar anecdotes of women's susceptibility to glare or their inferiority to men in matters of driving continued to appear for some time over the years.

In a number of instances, headlights were directly compared with women's breasts or nipples, a slang use that continues in some circles to the present day (figure 3.21).[63] Although no precise etymology is given in slang dictionaries, the usage dates to the early twentieth century and presumably stems from a sight that, for predominantly male drivers, was hard not to stare at. The analogy was only slightly veiled in a poster for an automobile exposition held in Geneva in 1906, which would likely have appeared risqué at the time, and appears from time to time in other representations of headlights. By 1940 it was familiar enough to slip into a Humphrey Bogart film, *They Drive by Night*. In one scene, Ann Sheridan plays a waitress behind the counter of a diner serving three truck drivers in the late night:

First Man: Nice chassis, huh, Joe?
Second Man: Classy chassis.
Waitress: Yeah, and it all belongs to me. I don't owe any payments on it.
Third Man: I'd be glad to finance it, baby.
Waitress: Who do you think you're kidding? You couldn't even pay for
the headlights.[64]

There is no shortage of objections from this period regarding the frailties of women drivers, even as automobile industry advocates ardently encouraged women to get behind the wheel. In time, however, the anecdotes evolved into science. In the mid-1930s a new malady arose: "tunnel vision … one of the worst visual deficiencies of drivers."[65] It was defined as a narrowing of the field of vision, with the sufferer unable to see more than between ten and forty degrees off a center of focus and prone to compensate by turning the eyes from side to side, away from oncoming traffic.[66] Attention might be more sharply focused in this condition, but the broader field of view was lost, leading to the later metaphorical use of the term to describe a narrow or uninflected perspective. Women, according to the largest relevant study at the time, were more likely to exhibit the deficiency, which seemed to add empirical weight to the claim that women "on the whole are not so expert as men" at driving. They were also harder to teach, the same study found.[67] A later study offered a "scientific viewpoint" to document female inferiority in night driving: following exposure to glare, men recovered normal vision an average of ten seconds faster than women. The investigators speculated that the reason was women's lower consumption of vitamin A, a factor in night blindness.[68] While headlights are not usually singled out in this literature, they nonetheless reproduce the conditions of narrowed vision. They can be considered in this sense as something of a prosthetic paradox. On the one hand, they remedy an inability to drive at night while, on the other hand, they do so by inducing a form of tunnel vision, a physiological defect and in this sense a kind of antiprosthesis.

THE SPEED MANIA IS ON THE WANE

I could see a car in their driveway so I stopped on the street below
just behind a petting party. Being a nice, dark, secluded spot,
it is an ideal place for such parties. The couple immediately moved
on about two blocks to escape my headlights.
Edith Johnson, 1934[69]

With headlights the focus of a national controversy over visual perception, calls rang out for more regulation and better lamps. "The headlights are a menace," began a typical comment, "they almost blind the pedestrian as the car approaches and make it impossible for the driver of a vehicle of any sort to see where he is going as he nears the dazzling lights."[70] Glare had long been a key problem of electric lighting; overcoming it was an explicit rationale behind the formation of the IES.[71] It had also been implicated in a range of ophthalmological maladies. Asthenopia, for example, was a term coined in the early nineteenth century to refer to ambiguous symptoms of eye strain such as headache, blurred, double, or generally diminished vision. By the early twentieth century it was linked to the concentrated brightness of electric light in particular, with automobile headlights a contributing factor.[72] With more and brighter headlights introduced over the following years, public clamor led to a "campaign against glaring headlights nation wide." Indeed, "war has been declared. … There can be no doubt that the glaring headlight problem is one of the most serious confronting the automobile engineers, the municipal and national authorities, and the public at large."[73] Whereas street lighting indicated a city's wealth, headlight regulation was soon seen as a measure of its civility.

There were no relevant federal or state laws governing automobiles at the time of their first appearance. Early regulations were municipal and usually concerned conflicts between automobiles and prior road occupants such as carriages, animals, bicyclists, and pedestrians.[74] Municipalities enjoyed considerable autonomy in setting headlight laws, which varied from one town to the next. Amber lens filters were acceptable in New Jersey but not in New York, which meant that drivers could be fined crossing the Hudson if they failed to change their lamps. Delaware had no restrictions on headlights but Wilmington did.[75] Newspapers frequently reported on "an owner who found his lights were regarded as all right in one place and all wrong in another."[76]

3.22
Pedestrians falsely believe they are well lit by bright headlights, whereas "to the driver the pedestrian illuminated by his headlights is just another object in the midst of a thousand other such vague shapes." Pedestrians were advised to wear light-colored clothing, to carry reflective objects, and to "consider yourself invisible and walk accordingly." Film stills, *When You Are a Pedestrian* (Oakland, CA: Progressive Pictures, 1948).

Courts heard cases regarding headlights under numerous circumstances: on curves, when the car was not in motion, at dawn and dusk; on whether the dashboard constituted the "front" of the car where the headlight should be mounted; on who was at fault when a car struck an unlit vehicle or bicycle; even on whether it was legal to drive at night on unfamiliar roads. Failing to pull over when faced with bright oncoming lights could be illegal: "It is negligence to drive on when one's eyes are blinded by the headlight of an approaching vehicle."[77] Other courts considered the same act "gross negligence," a deliberate rather than an inadvertent failure to exercise proper care. In such cases, courts had reason to hold drivers to "strict accountability" even when pedestrians behaved erratically (figure 3.22).[78] Glare from the sun and light distorted by fog or rain were likewise reasons to stop the car.

As car sales skyrocketed and more drivers explored both city and countryside by night, cries for better standards and tighter regulation regarding automobile manufacture, road quality, and headlights reached a new high. One result was the Federal Aid Road Act of 1916, signed by Woodrow Wilson and requiring localities to adhere to federal guidelines, which made roadbuilding a national priority.[79] Another result was a surge in legal proceedings regarding headlights. Dockets swelled with cases.[80] In many instances, however, the prosecution foundered. Early laws were notoriously vague, usually specifying only that headlights be "non-glare," just as speed was often limited to what was "reasonable and proper." But glare was especially hard to define. In response to a Massachusetts headlight law, the president of the state's Automobile Club complained: "The police tell me they don't know what the law means, they are uncertain what constitutes glare."[81] Enforcement was challenging, and police often ignored headlight laws unless a recent accident had renewed public outcry. "Many amusing stories" were told of police ineptitude in judging glare when, for example, one man was arrested twice one night, "once for too much light and again for not enough." The man, it seems, rectified his first violation by putting some rags over his headlights, whereupon a second officer stopped him. The driver explained what had happened and asked the officer "to tell me just how much light I could use. He tried to explain the matter to me, but without success."[82] A Texas law was voided by the courts for "indefiniteness; the glare and brilliancy not being properly defined." For the citizen eager to obey the law, "there must be some certain standard by which a person can determine in advance whether or not he is complying with the statute."[83] J. B. Replogle, an engineer with an Indiana-based electrical supplier for automobiles, explained how he had contacted fifty-eight municipalities across the country and found their regulations "futile and in many instances ridiculous." The underlying reason, he said, was the difficulty in defining glare.[84]

There Are Two Sides to Night Driving

3.23

Unreasonable "sons of glare" corrupt the visual field in their "blazing chariots" and disturb the peace of the countryside. "There are Two Sides to Night Driving," *Light* 8, no. 6 (Summer 1930): 18. Reproduced with permission from the Archives and Research Center, Museum of Innovation and Science, Schenectady, New York.

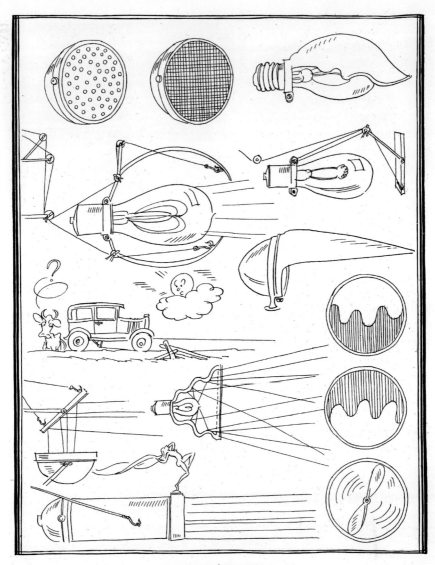

They Shall Not Glare!

3.24

Parody of the distended market for antiglare devices. "They Shall Not Glare," *Light* 8, no. 6 (Summer 1930): 21. Reproduced with permission from the Archives and Research Center, Museum of Innovation and Science, Schenectady, New York.

Setting standards for glare meant balancing two opposed needs: more light on the road for those inside the car and less light on the road for those outside the car. As the press characterized the dilemma: "The confusion incident to the problem of regulating glaring headlights ... is caused by the attempt to discover a method or means for the absolute elimination of glare without reducing to any appreciable extent the road illumination."[85] In addition to the optical properties of the headlamp, engineers had also to consider the speed of the vehicle and the effect of "distracters," that is, other stimuli that might divert cognitive time away from visual perception. Most important, glare was in the eye of the beholder, it was "affected by age, darkness adaptation, brightness of background, contrast between that and test object, duration of exposure, brightness of pre-exposure, color of stimulus, portion of retina stimulated, movement of test object, flicker, disease, and fatigue."[86] Sensitivity to glare varied widely even among perfectly healthy individuals, as the IES found in the largest-scale study it had conducted as of 1918.[87] Individuals recovered their night vision at different rates in relation not only to the light source but also to a particular stage of dark adaptation. The eye recovered quickly from a "relatively blinded state," for instance, but complete adaptation could take an hour.[88] In other words, in their efforts to measure and thereby regulate safety on public roads, states had first to understand how individuals learned to navigate the subjective as well as the objective aspects of the space of the headlight.

3.25
Conflict of headlights, one law-abiding, the other aggressive, dangerous, and illegal. Advertisement, Legalite Lenses, Hyslop Bros., *Hardware and Metal* 29, no. 22 (June 2, 1917): 36.

In response, drivers took it upon themselves to paint their headlamps, slip colored gels over them, wrap them in cheesecloth and rags, or bend the mounting bracket so the lamp would point downward, and they would carry kits with buckets, pigment, fabric, string, and hand tools to do so.[89] They mounted lights high on the car so that they shone downward rather than straight forward, which had the added effect of a more natural illumination.[90] They tried elaborate headlamp-masking schemes whereby light would be thrown only to the right side of the road, which avoided blinding others but produced "at first an uncanny" sensation owing to the asymmetrical contrasts created by sidewise lighting.[91] They also tried a variety of spectacles and goggles designed to minimize glare.[92] They bought "patent" devices from supply houses that sold products from freelance inventors as well as established manufacturers. But even these proved difficult to regulate (figures 3.23, 3.24, 3.25, and 3.26). The inventor of one dimmer had adjusted it for a driver just minutes before the driver was pulled over for excessive glare. Luckily for the driver, the device came with a $100 guarantee, which was paid once the court found the inventor guilty of inadequately adjusting the headlight. Other defendants acting in good faith purchased similar devices that were subsequently found to be faulty, leading to their being fined for excessive glare.[93]

Instead of these temporizing measures, what was most needed, Mitchell May maintained, was a mass-produced technical solution, some sort of "glare-subduer in the construction of the lamp itself."[94] Headlamp manufacturers were happy to oblige. Headlamps were already a lucrative market: they constituted one quarter of General Electric's lamp sales in 1924, for example, and cars averaged six headlamp replacements per year. Competition for headlamp dollars resulted in numerous lamps and lighting arrangements and frequent, often confusing enhancements to filaments, switching circuits, reflector shapes, lens geometry, and so on.[95] Bulbs were screwed, snapped, or slid into place, which could throw off the lamp's focus, as might bumps in the road. Further, lamps were sometimes stolen, leading to additional gadgets to make their removal time-consuming.[96] Trade periodicals and newspapers discussed ingenious solutions, such as a French design consisting of a

3.26
"No greater nervous tension and actual fright can be packed into a few seconds at night than that occasioned on an unseen road by the blinding glare of a car in the hands of a reckless driver" or "headlight hog," a driver turned bestial by powerful headlights. These glasses are darkened in such a way that drivers need only turn their head slightly to have the glare nullified. "Making the Night Safe for Automobile Drivers," *Popular Science Monthly* 93 (October 1918): 40.

strip of twelve bulbs running along the top of the windshield. The road would be evenly lit while the intensity of the source was dispersed, and the device, mounted on brackets, could swivel backward to light the car interior when needed or when the driver needed to open the hood.[97] To assuage the problem, citizens petitioned city and state government for lists of approved lamps; some of these include more than two dozen types.

Makers also tried various dimming schemes such as tilting lamps downward to "dip" the headlight, which entailed more moving parts subject to deviations from time, turbulence, and neglect. Still, as General Motors claimed for its Cadillac line, dipping the lamps avoided "derangement of the vision" and assured "safe and pleasurable night driving."[98] Part of the problem was that car batteries were not yet standardized, especially for all-electric vehicles, which could have one of four different types that output a range of voltages, each requiring a different type of lamp.[99] Though manufacturers had "flooded the market with various devices, each guaranteed to give perfect driving light and eliminate the glare," they were reluctant to take ultimate responsibility: they "apparently feel that the problem does not in any way belong to them."[100] Often, manufacturers laid the burden of care on consumers. "I Have *Only Myself* to Blame," ran an advertisement for Osgood lenses, with a prismatic lens that threw light downward. The ad highlighted the driver's remorse at having chosen an inferior brand of headlight that "blinded and frightened the other man. … It was an awful nightmare." Although safety was the consumer's responsibility, speed was his reward. The driver voiced remorse in this and similar advertisements while the manufacturer pitched excitement and euphoria: Osgood "lights the road far ahead for greatest speed" but without the distraction of dimming or dipping (figure 3.27).[101]

3.27
Remorse and blame for violating visual protocols with glaring headlights that blinded another driver. "It's all an awful nightmare," Osgood advertisement, *Chicago Daily Tribune*, Aug. 6, 1916, 8.

Headlight laws gradually became more uniform. In 1918, New York State made national news by consulting illuminating engineers to determine measurable standards. The resulting law, however, in specifying candlepower at a precise point in space relative to the headlamp, meant police had to carry specialized instruments to determine compliance.[102] The steady march toward standards made night driving less stressful, but so too did empathy for the effects of glare on other drivers. As the *Atlanta Constitution* reported, "The speed mania is on the wane," signaling the return to a general sense of civility as well as more consistent enforcement. Even railroads stopped racing to shave minutes from schedules. The whole country, it appeared, was calming down.[103] Professional and political opinions eventually converged on the need for uniformly lighted highways.[104] And a number of years would pass before state and federal lawmakers, illuminating engineers, automobile clubs, headlamp makers, automobile manufacturers, and the driving public would settle on national standards for headlights, most decisively in the late 1930s with the arrival of the sealed beam headlamp, an integrated assembly of lamp, reflector, and lens within a sealed enclosure filled with an inert gas. All elements were optimized in the factory and rigidly mounted by the manufacturer. Drivers could stop thinking about their headlights and instead focus on the luminous kaleidoscope ceaselessly unfolding before them (figure 3.28).[105]

3.28

The headlight is less of a harrowing place to be after the sealed beam headlamp becomes a commonplace. "Only last summer I was scared all the time" while driving after dark. After Tom had new sealed beam headlamps installed, "it all seems so simple ... so clear ... so sure." Headlight advertisement, General Electric, *Life Magazine*, March 18, 1940, 93. Ellipses in original. Reproduced with permission from the Archives and Research Center, Museum of Innovation and Science, Schenectady, New York.

CONCLUSION

NIGHT DRIVING IN NEW YORK. Bright light blurred in the distance. ... The dampness is refreshing. Smell of escaping gas. Bumps and sudden stops. Torn-up roads. Street-repair machinery looking like skeletons of antediluvian animals. Again a blurred and garish sign. Nearing it I could see that it mingled with the moonlight on the river. ... A patch of light with a crowd around it. An accident? No, a tricky-links golf tournament. Jouncing and speedy brake work. ... Then Riverside Drive. The bridge. There is a work of Men. There is Beauty.
Lavinia Riker Davis, 1933[106]

Like cinema, the other great novel visual experience of the early twentieth century, night driving was an alloy of visual flux and bodily stasis organized by a projected beam of horizontal light parallel to and nearly coincident with the viewer's eye. Moviegoer and night driver alike gazed forward at a succession of scenes, serial tableaux melting one into the other to form a shifting landscape of light reflected from surfaces. The two experiences converged nearly from the start. Strapping himself in 1868 to the cowcatcher of a train, beneath the headlight whose "strong glare" made him invisible to flag-men at street crossings, one man described the experience as a sequence of vivid scenes set off from one another by semicolons:

> The cone of light from the dazzling lens above me with then cut the darkness in its onward rush with startling clearness. Now flashing up the rocky sides of some mountain gorge, illuminating rock, tree and shrub with almost daylight distinctness; anon losing itself in the surrounding darkness as we emerged into the open country beyond; then shooting along the rails ahead, making them look like glistening threads until they disappeared in the darkness beyond.[107]

Rare at the time he wrote, the author's visual experience was later replicated in a cinematic trope, "the phantom ride," made by placing a camera on the front of a moving train.[108] And, of course, it was multiplied countless times in night-driving cars. But however much they were alike, cinema and night driving involved considerably different stakes, not least in terms of the significant social negotiations necessary to introduce and manage millions of serial luminous easements on public roads at night.

Unlike the steady cinema screen, the landscape beyond the car's windshield was populated and alive. The driver's visual field could be struck by glare, passing blind spots, flashes, an occasional "dancing 'will-o'-the-wisp' effect" caused by the flickering spokes of a bicycle, and other unresolved luminous features. Weather conditions such as snow, fog, and rain could throw "a dazzling veil" over the eyes. Light from oncoming traffic was "caught and broken on a wind shield" or the driver's goggles. Everything was constantly coming into form and then passing by.

Even the 1868 train rider described feeling fearful when an approaching headlight suddenly glared at him, although he had tethered himself to the train for safety. Moreover, automobile drivers had to anticipate the visual field of other drivers even as the means of communicating with those other drivers was stripped to a minimum. That is to say, night driving presumed a unique kind of intersubjectivity consisting of both human perception and mechanical capabilities.[109] Pedestrians, for their part, were warned to "understand how the approaching flood of light makes their footing uncertain by the deepening and dancing of shadows, how the nearing light draws the eye and even checks movement as in a kind of momentary fascination"—that is, it causes paralysis.[110] Simply to walk along or beside a road at night was to enter a new world where recurrent roving rays led to beguiling misperceptions that could throw off one's balance, with ruinous results.

While drivers, like moviegoers, stared into a mise-en-scène of modernity, constantly unfolding scenarios of staggered, staccato reflections interwoven with snaking shadows, they were also a part of it. The light cast by the headlamp reconfigured the appearance of objects and, as a result, recast the spatial knowledge needed to move forward. By day, light fell from above, equally on all things; drivers calibrated their motion against a diverse visual backdrop uniformly rendered in all directions. By night, the headlamp projected a bright, horizontal light. Objects appeared less as solid forms than as formations of variously reflective surfaces, which changed as the car moved and the angle of the projecting beam changed. Even small but shiny things outside the cone of the headlight could loom large in the visual register. Thus, for the night driver, the world seen through the lighted tunnel of road ahead became a specular rather than a geometric construct.

Drivers were repeatedly reminded that they inhabited a technological, regulated, unsteady, and, often, dangerous space.[111] They kept abreast of local laws; knew how to change a flat tire, replace headlamp bulbs, or refuel acetylene tanks; knew how to add motor oil, clean carburetors, and manage many other frequent repairs. Since many roads were poorly paved or entirely unpaved, headlamps frequently went out of whack (figures 3.29 and 3.30). Adjusting them was a chore. To readjust the lamps, drivers were advised to place their car ten or fifteen feet from a wall and move the lamp bulbs forward or backward in relation to the reflector to generate as condensed and uniform a luminous shape as possible.[112] Following this operation, the car was to be moved to level ground, the driver walking seventy-five feet from it to place "a mark on a board, broom handle, or on the coat of an individual 42 inches from the ground" to confirm that the beam was neither too high nor too low.[113] Moreover, different manufacturers provided different means of making adjustments, sometimes involving an adjusting screw on the external casing, the easiest method; sometimes requiring a driver to remove the lens and twist the bulb directly; and sometimes requiring the removal of the reflector so that internal mounting screws could be loosened and the bulb moved and then retightened, the most time-consuming of options.[114] Even buying bulbs required some technical skill; savvy drivers knew to check the integrity of the filament and solder joints since manufacturing quality varied. As late as 1925, the Los Angeles County Highway Motor Patrol and the California Automobile Trade Association conducted "headlight adjusting classes" for drivers, led by hundreds of certified "motor-car headlight adjusters."[115] Although such repairs were troublesome, mechanical awareness allegedly brought satisfactions and contributed to the driver's "complement of joy" in motoring.[116]

Rather than imagined as an abstract quantity or a set of passing images, speed at night was explicitly registered as the product of vehicle quality, headlamp maintenance, legal codes, road conditions, physiological training, the fluid character of the visual field, and the driver's ability to interpret it in a hybrid state of aestheticized visual vigilance. As Gibson and Crooks noted, the automobile is a spatial and somatic field that yields kinesthetic, auditory, tactile, and visual impressions, as well as corporeal sensations, such as a sense of "traction," transmitted from the roadway through the car's body, which, in the early days of automobile suspensions, could be considerable and absorbed somatically by the driver, who would need to interpret rumbles, trembles, shudders, and shivers for their bearing on continued forward motion. Driving, they concluded, "is a matter of living up to the psychological laws of locomotion in a spatial field."[117] With the automobile's occupants thus acutely aware of technological and legal conditions, night driving was a profoundly embodied experience.

More than simply knowing how to maintain their car and lamps and to follow the law, night drivers could feel themselves projected into the space of the headlight, seemingly generating their field of view rather than passively receiving fleeting images through a viewing device. Drivers focused forward, repeatedly scanned a constantly shifting field of view to classify heterogeneous luminous signals in terms of hindrance, continuance, or irrelevance, and all at relatively high speeds. Numerous publications offered tips for judging whether one or another of these points and surfaces of light might be an obstacle well in advance of identifying it as an object, thereby adding to the general sense that the space of the headlight was physiologically active and had profound bodily consequences. With the driver's attention perpetually placed well forward of the body, the headlight rested on a predictive premise, not just lighting the road ahead but anticipating movement forward. In this way, with the driver continuously following it, the headlight, nearly coincident with his gaze, seemed to anticipate his motion, as if light were projected as much by the pressure of the body moving forward as by the headlamp itself.

3.29
The space of the headlight stimulated all the senses, not just vision. Headlight advertisement, General Electric Co., *Saturday Evening Post*, Aug. 1, 1925. Reproduced with permission from the Archives and Research Center, Museum of Innovation and Science, Schenectady, New York.

3.30
The headlamp is a sensitive device that needs repair and at the same time facilitates other nighttime automobile repairs. *A Lamp in Reserve*, desk blotter no. 220, General Electric Co. ca. 1910s. Reproduced with permission from the Archives and Research Center, Museum of Innovation and Science, Schenectady, New York.

3.31
A woman driving into the forever receding space of the headlight. *Man and the Motor Car*, ed. Albert W. Whitney (Hartford: Department of Motor Vehicles, State of Connecticut, 1940), 133.

Directing attention and forecasting direction, the headlight appeared to many drivers as an emanation from themselves, a motorized ocular extramission. As one account put it, the modernized "autoist unconsciously follows the beam of light in driving," as if it were an ancient instinct (figure 3.31).[118] Night drivers registered the mechanical energy of the car as a shifting specular world generated by their own forward motion. In turn, as the car's propulsive power was internalized as an extension of the body's own motor capacity, so the headlight was experienced as an almost tangible projection of the power to see, a self-produced portrayal of the seeing self. The headlight was a representation, in other words, of self-propulsion and at the same time actively shaped a lived reality for the driver and passengers and whomever strayed within its luminous cavity. It was as much a tool to navigate the world as to observe it. These many facets of auto-specularity suffused the night drive, one of the most exhilarating individual experiences in the early twentieth century.

When the historian Sigfried Giedion said in 1941 that highway driving was the best means of experiencing the space-time continuum of modernity, he was only half right. For Giedion, modern space was multivalent and changeful. "In order to grasp the true nature of space the observer must project himself through it," he wrote, and driving "up and down hills, beneath overpasses, ramps, and over giant bridges" was self-projection through space at its most vivid.[119] In this formulation, space was made modern by mechanical motion across a landscape of massive infra-structure. Taking the headlight into account, however, makes clear that the driving self, not just the space, was likewise made modern. Learning to drive at night required the acquisition of new perceptual habits that, in time, became naturalized and all but lost to conscious attention. In Gibson and Crooks's terms, drivers internalized "semi-automatic perceptual habits" rather than stiffening themselves in "a continual state of strained attention."[120]

To position oneself at the vertex of a cone of light and propel it across a darkened landscape must count as one of the most startling visual experiences of the twentieth century. Likewise, to stand in the path of an oncoming headlight was certainly one of the most frightening of everyday occurrences. Only after a great deal of debate regarding their creation of a new kind of psycho-mechanical space, as well as their disruption of prior spaces, were night-driving cars granted a rolling easement to serially occupy the night. Incorporating a space that was thrilling, dangerous, and unprecedented in its rapid diffusion—shaping it, navigating it, avoiding it, adjudicating it, learning to live with it—into the social fabric of routine life surely was a remarkable historical episode, however much it has been so naturalized that our routine encounters with it seem as if they could hardly have been otherwise.

4
LIGHTING FOR LABOR

INTRODUCTION

Industrial managers are coming to realize that light not only increases output and decreases accidents, but also plays a most important part in making the plant a cheerful and pleasant place in which to work.
C. W. Price, 1920[1]

Steady, bright light is conducive to work (figure 4.1). At first glance, it is hard to see this claim as anything more than common sense based on ordinary experience. On closer examination, the truism turns out to have a history. It was formulated by capitalist reason in factories as a product of lighting innovations, first gas in the early nineteenth century and then electric in the late nineteenth and early twentieth centuries. As sites of concentrated labor, factories—textile mills in particular—were large users of artificial illumination and early adopters of lighting technologies. They were also privately owned, organized for profit, and had only a few constituencies: workers, management, and, in the context of lighting, illuminating engineers. These finite aims and few actors considerably narrowed the rationale for lighting and simplified its historical politics. Questions of privacy, leisure, sociability, and so on were eclipsed by the singular goal of economically enhancing output. Lighting was conceived in the factory setting not as a public good or service but as a factor of production.

While adequate illumination had long been a central concern for workshops generally, the adoption of electric lighting, especially in the context of theories of scientific management and concern for efficiency, led to a new attitude toward vision's role in industrial production. Worker safety and supervision were paramount, but illuminating engineers increasingly isolated seeing as a form of what came to be called "visual labor." They experimented with the intensity and spatial distribution of light, studied its effect on workers' efficiency, devised lighting protocols for different types of seeing, developed economic arguments regarding lighting's costs and benefits, and staked claims regarding the morality of good lighting. In essence, industrial experts tried to Taylorize vision, to treat vision as a commodity in order to assimilate it to the productive rationality of the factory system, to concentrate its utility and subject it to ever finer divisions of time. The architectural setting became in this way a laboratory for modernizing vision, a venue in which to analyze lighting's productive potential and find ways to accelerate seeing and make it more efficient.

With the luminous environment understood as a contributing factor of productive labor, other spaces of social production came to be scrutinized as well. Schools in particular were seen as workplaces for the young, where teachers

A RIGHT WAY AND A WRONG WAY TO USE LIGHT

IN THE PICTURE AT THE LEFT, THE WORKMAN'S EYES ARE SHADED AND ALL THE LIGHT IS CONCENTRATED ON HIS WORK; IN THE PICTURE AT THE RIGHT, THE WORKMAN IS BLINDED BY THE GLARE WHICH NOT ONLY IRRITATES HIS EYES BUT MAKES IT HARD FOR HIM TO SEE WHAT HE IS DOING

4.1

A right light and a wrong light for work: light on the work at hand, or light in the eyes, which makes objects much harder to see. Clara Brown Lyman, "The New Illumination," *World's Work* 28, no. 2 (June 1914): 153.

managed the efficient production of literate and industrious citizens. With "light conditioning recipes," engineers proposed lighting schemes for homes too, deeming domestic tasks productive activities. Even domestic leisure and rest might be optimized in terms of lighting. Conceived as places that produced sales, shops too were recalibrated in terms of their lighting. Factories, in short, were incubators for ideas of productive vision that were assimilated to a wide range of other spaces as an imperative element in the visual infrastructure of labor.

DIMLY LIT LABOR

The good huswive's candle never goeth out.
William Baldwin, 1584[2]

Throughout history, most things were made during the day. The sun's rising and falling powerfully affects human physiology, making night nearly everywhere a time of rest. However, this natural law is ramified in cultural ways. Noting mistakes made in dim light, European craft organizations prohibited nighttime work in order to control the quality of goods, a proscription common among trades requiring detail work, such as making lace or working gold. Daytime labor was also more conducive to inspection by master craftsmen, by purchasers, and by tax collectors. Further, devils and demons were thought to prowl by night, making all sorts of mischief. In general, with vision compromised by darkness and alertness blunted by fatigue, night was a perilous realm best navigated with sleep (figure 4.2). There were exceptions, however. Medieval farmers, for instance, worked through the night during harvesttime under the "harvest moon" that followed northern Europe's growing season.[3] In towns, bakers worked during the wee hours to prepare bread for the morning trade. Denizens of the dark, able to adulterate flour unsupervised, they endured shady reputations colored by common suspicions about the night.[4] Likewise, trades that required steady fires, such as iron smelting, glassmaking, and beer brewing, kept tenders active through the night. Tailors and cobblers could be given explicit license to work nights to fill orders for elite families. Certain services, such as removing night soil and carting away corpses, were more efficient and sanitary when conducted at night (figure 4.3).[5] Finally, many people had too much to do while others just procrastinated. As the English minister John Clayton wrote in 1755: "The temptation to lie in an extra hour in the morning pushed work into the evening, candle-lit hours."[6]

Women in particular would work at night. Laundry was especially time-consuming, whether it was done for one's family or as a servant for others or taken on for extra income, compelling the conscientious housekeeper to stay up late. Nightworking women might gather at one house to keep company and to share the

4.2
Some work remained undone at nightfall, and workers strove to remain attentive and complete their chores. They were not always able to resist the nighttime's tide of sleep.
Sleeping Girl with Needlework in her Lap, mezzotint, Gerard Valck, n.d.

4.3
Social values dissolve at night, but some forms of labor, such as shaving a customer, visible through the window in the lower left of the print, continue regardless of available lighting.
Four Times of the Day: Night, engraving, William Hogarth, 1737.

artificial light.[7] Lace makers even came up with their own lamp, made from a jar of water, often tinted light blue and placed before a candle to refract and concentrate the light. The resulting sideways beam sharpened contrasts and made individual threads easier to see.[8] Beginning in the fourteenth century, women also gathered and worked late to spin, card, and weave cotton, wool, and linen as part of the "putting-out" system of textile production.[9]

Lighting technologies changed little until the eighteenth century. Rushlights, made by pulling a common plant through rendered animal fat, had been used since antiquity. They were simple and cheap and, importantly, not taxed.[10] Pinewood, rich in flammable tars, or small burning piles of peat could also illuminate nighttime work. The phrase "burning the midnight oil" linked night work to oil lamps, also ancient in origin. Tallow candles were a boon for those who worked at night and at the same time a symbol of drudgery. As the Elizabethan writer Thomas Dekker asked a candle in 1606:

> How many poore *Handy-craftes* men by
> Thee have earned the best part of their living?[11]

However useful, such light sources would appear dim to modern eyes and brought inconveniences that would be intolerable today. Lamp oil was often impure and smelled wretched when burned. Candles were usually badly made and hardened with time. They sputtered and guttered, exuding noxious odors and occasional sparks. Wicks needed constant care. One report, admittedly biased against candles, held that the common candle must "be snuffed about forty-five times during the time it is burning" to get rid of carbonized residue that reduced its efficiency.[12] Candles were also costly. The expression "not worth the candle" captured the relative expense of artificial lighting in relation to the value added in the making of new goods.[13] Made from animal fat, candles were sometimes eaten in lean times, when food was more valuable than light.[14] Moreover, igniting any light source at all, usually in the dark, was a protracted affair before the invention of friction matches.[15] In short, sources of artificial light for night work were expensive, exacting to kindle and maintain, malodorous, and, for all the bother, shed only a dim, wavering light that was wearying to work with.

MANUFACTURING VISION

Vision is an indispensable tool in modern industry.
Allen Kiefer Gaetjens and Dean M. Warren, 1939[16]

The emergence of the factory system in the eighteenth century consolidated labor within a new building type and, as a corollary, intensified demand for artificial lighting. Buildings housing machinery and workers and dedicated to production have been known since antiquity, but they were few and limited in scope. Though a modern silk throwing mill was built in Derby in 1719, William Fairbairn, a multi-talented Scottish civil engineer, held the view that as late as 1784 "there were no factories, properly so to speak."[17] At that time, entrepreneurs began to exploit the steam engine, new supplies of cotton, and new building materials such as iron to assemble greater numbers of workers to operate centrally powered machinery to perform incremental tasks in a larger production process. The textile industry was the first to take broad advantage of the efficiencies introduced by factory production.[18] The division of labor noted by Adam Smith, among others, dramatically increased production and lowered costs, spurring further innovations to meet the consequent rise in market demand. In assigning workers to a limited set of actions on specific machines, ever finer divisions of labor also ended up leading to evermore precise accommodations in the factory for physical—and visual—labor.

The Trouble with Daylight

It was shown that an extremely large percentage of the daytime, when one would expect complete reliance to be placed on daylight, actually required the use of artificial light, this being especially true in the older plants with restricted window surface. ... In the dark months, such as December, for instance, it is found that the lights will be needed for practically all the daylight hours in order to secure the intensity that is needed for proper workmanship.
***Electrical Review*, 1918**[19]

Early mill buildings tended to be long, narrow, and multistoried in response to the peculiar geometry of their power drive, a system of shafts and belts that distributed energy from a central water wheel or steam engine to an array of individual machines.[20] Narrowness also maximized the penetration of daylight into the building's interior (figure 4.4). Fine or precise work was placed close to the windows, while the building's dim, central spine served storage and circulation. Thus mills were organized in bands parallel to exterior walls and graduated according to workers' visual needs.[21] Although effective, the belt-and-shaft system, as it came to be known, presented some problems, including the need for constant maintenance; an unctuous atmosphere composed of warmed grease rising from constantly lubricated belts and the shafts that turned in them and, especially in the case of cotton, flammable lint; the need to set the entire system in motion even when only a few machines were in use; and a sequence of manufacturing operations organized around the delivery of power rather than the logic of assembly.[22]

Daylight, for its part, was also highly effective, being bright, abundant, free, and offering familiar or natural color rendition. But it too had problems. For one, its intensity diminishes quickly indoors, linearly decreasing with increasing distance from the window. Windows were limited to about 30 to 35 percent of exterior wall area in order to maintain load-bearing walls strong enough to support the heavy power drive system and all the machinery.[23] But even the largest windows could be blocked by the power drive or other equipment, especially as new elements were added to existing setups. As late as 1911, a factory inspector observed that "in many instances where natural light is used, machinery is so placed as to obstruct this light, and thus the duties of the workers must be performed in semi-darkness," with detrimental effects on the quality of the work and the health and safety of the worker.[24]

Moreover, free daylight was hardly free. It had to be modulated by awnings, shades, screens, or diffusing treatments on glass. It was most effective with high ceilings, tall windows, or roof monitors, all of which increased construction and maintenance costs.[25] Daylight also varied with the time of day, the time of year, the orientation of the building, the surroundings, passing clouds, and the weather, not to mention interior conditions such as obstructions or dirty finishes (figures 4.5 and 4.6). As a result, daylight resisted quantification at a time of rising precision in capital accounting practices. Even into the twentieth century, illuminating engineers readily acknowledged that they measured artificial lighting far more easily than natural lighting, leading many to favor the former, a preference that led to entirely windowless factories in the 1930s.[26] Engineers, along with factory managers, came to believe that the vagaries of daylighting led to waste and work of varying quality. Winter batches, for example, made when days were shorter and the sun was lower in the sky, were poorer in quality than those made in other seasons.[27] For industries beginning to standardize operations as a means of maximizing output while reducing costs, such discrepancies were anathema.

4.4

Mill No. 1, built in 1835 in Lowell, Massachusetts, was 45 feet, 6 inches wide and 139 feet long. Its length was more than doubled that same year to 375 feet, while the width remained the same. Lighting was provided by windows evenly spaced along its two-foot-thick brick walls. Daylight fell off quickly with increasing distance from the windows. *Southwest Elevation—Mill #1, First Floor—Mill #1, Boott Cotton Mills, John Street at Merrimack River, Lowell, Middlesex County, MA.* Image courtesy of the HABS/HAER/HALS Collection, Library of Congress, Prints and Photographs Division (http://www.loc.gov/pictures/item/ma1289.sheet.00004a).

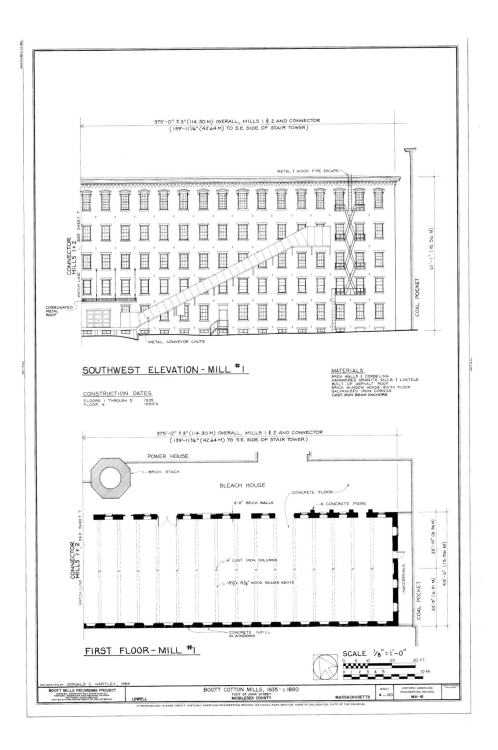

LIGHTING FOR LABOR

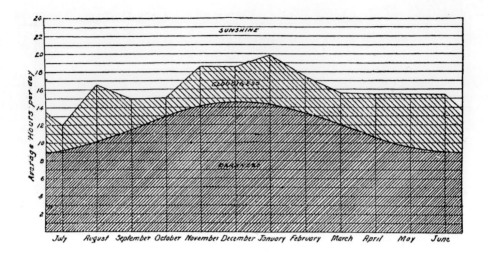

4.5

The vagaries of daylighting, evident in this chart depicting lighting levels across the seasons, ran counter to a growing need for predictability in the factors of production. Leon Gaster, "The Economic and Hygienic Value of Good Illumination," *Journal of the Royal Society of Arts* 61, no. 3142 (Feb. 7, 1913): 308.

4.6

Even in the early twentieth century, machinery, along with other obstructions, could block the daylight available to laborers. *A Little Spinner in the Mollahan Mills, Newberry, S.C.*, Lewis Hine, photographer, 1908. Library of Congress, National Child Labor Committee Collection, Prints and Photographs Division, (http://www.loc.gov/pictures/item/ncl2004001280/PP).

Gas Light

Even the ideas of day and night, of rustic simplicity
in the old statutes, became so confused that
an English judge, as late as 1860, needed a quite
Talmudic sagacity to explain "judicially" what was day
and what was night.
Karl Marx, 1887[28]

In the late eighteenth century, mills, striving to maximize output from considerable capital investment, experimented with evening and night shifts and, in the process, became major consumers of lamp oil and candles and certain newly available lighting technologies. Richard Arkwright's 1771 Cromford Mill in Derbyshire operated around the clock: a night shift prepared materials for the spinners, who worked by day. As a 1790 visitor wrote, "The mills never cease off working. … Seven storeys high and fill'd with inhabitants [they] remind me of a first rate man-of-war and when they are lighted up on a dark night look most luminously beautiful" (figure 4.7).[29] In 1806 a Manchester, UK, spinning mill burned on average 1,500 candles for eight hours each night for twenty-five weeks; a Manchester cotton mill kept ten-hour days in the winter by burning over 1,000 candles a night.[30] Evening hours usually began in mid-September, marked by the first "lighting up," with a final "blowing out" around mid-March that workers came to celebrate.[31]

Mill owners in the English midlands seized on gas lighting as it became available in the early nineteenth century.[32] Gas lamps had many benefits: they burned relatively cleanly; they did not spark; they required little maintenance compared with candles or oil lamps; they could not spill oil or drip tallow; and they were easily extinguished. Moreover, mill owners could choose from among dozens of burner types to suit virtually any purpose.[33] Of particular importance, insurance companies strongly approved of gaslight and welcomed back clients after having cut coverage because of the numerous fire-related losses common to textile mills, which had dozens or hundreds of open flames burning in a greasy and lint-filled atmosphere.[34] The greatest benefit of gas was the light itself, which was brighter and steadier than light from other sources and more uniform from lamp to lamp.[35] Exactly how bright was a live question as there were no objective measures yet, but it was generally judged to be from six to sixteen times as bright as a candle and at least three times brighter than the best oil lamp.[36] Arrayed in rows to light complex machines operated by ranks of diligent workers within a new building type that served novel productive processes, they were a beacon of progress (figure 4.8). Edward Baines, a newspaper editor and owner and a liberal Member of Parliament, described such a scene in his 1835 survey of British cotton manufacturing:

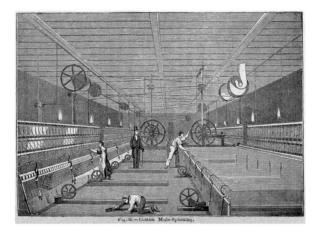

4.7

A thousand candles, or oil lamps, glow with a light to rival the Moon. *Arkwright's Cotton Mills by Night*, Joseph Wright, ca. 1782–1783. Private collection.

4.8

Cotton Mule-Spinning. The print suggests adequate and remarkably even lighting in the mill interior from its wall-mounted gas lamps, but this was very likely not the case in reality. Andrew Ure, *Philosophy of Manufacturing* (London: Chas. Knight, 1835), 308.

> At the approach of darkness the building is illuminated by jets of flame, whose brilliance mimics the light of day. ... It must be acknowledged that the cotton mill presents the most striking example of the dominion obtained by human science over the powers of nature, of which modern times can boast.[37]

At the same time, gas introduced new inconveniences: poor-quality gas could have a terrible odor, and, while steadier than candlelight, light from gas sources still flickered. Although light from gas was brighter than light from other sources, every task required at least one lamp. More workers meant proportionally more lamps. As a result, ambient air temperature rose. One mill reported in 1883 that 456 looms lighted by 457 gas burners had raised the workshop temperature 25 degrees within an hour.[38] Temperatures at another mill rose to 90 degrees by the end of a shift.[39] Many factories installed fans to ameliorate the heat gain, but this led to even more distracting flicker.[40] While gas made night work more practicable, night shifts did not become commonplace as a result of the new technology, at least in textile mills.[41] Many mill owners noticed that, despite the comparatively brighter light, workers continued to make more mistakes at night than by day.[42] Burning coal gas vitiated the air, leaving workers light-headed, a condition that came to be called "mill fever" and was brought about by "the pestiferous atmosphere produced by so many breathing in a confined place, together with the heat and the exhalations of grease and oil. All these causes are aggravated in the winter time by the immense destruction of pure air by the gas that is needed to light up the e[s]tablishment." The condition was debilitating rather than deadly, but "unless the constitution be very strong, leaves its pale impress for life."[43] Nonetheless, gas lighting continued to be used in a number of textile mills well into the first decades of the twentieth century.

Electric Light

The first use of electricity for interior illumination on any large scale came with the introduction of arc lights among the textile mills of New England, and the operation of these novel lighting systems was immediately followed by a number of fires. People were very much surprised.
C. M. Goddard, 1922[44]

Just as with gas, textile mills were among the first buildings to introduce electric lighting systems.[45] Factories experimented with arc lamps beginning with Humphry Davy's first demonstration of them, but they only became feasible in the 1870s with dynamos and lamps that burned more steadily. By 1881 the engineer and entrepreneur Charles F. Brush had more than 6,000 carbon arc lights in operation, of which a plurality, about 20 percent, were installed in textile mills, with another 19 percent used in other types of factories.[46] The lamps were excessively hot and often sparked. Their installation was followed by a rash of fires, with one insurer enduring a 35 percent claims rate.[47] The Edison Company proposed supplementing and even replacing ambient arc lighting with task lighting by incandescent light, its specialty.[48] Charles Jeptha Hill Woodbury, a mechanical engineer by training who spent much of his career working for insurance companies, noted in 1882 that recent improvements such as incandescence made it "feasible to bring electric lighting from the laboratory to the commercial world, creating an element in manufacturing affairs."[49] Although gas remained a viable source of light thanks to Carl Auer von Welsbach's patent for a durable mantle that illuminated more brightly than an open flame, electricity had shoved gas off the trajectory of material advance. In the words of Allen Ripley Foote, an advocate for municipal ownership of utilities, the adoption of electric lighting over gas "is a question of Progress *versus* Retrogression."[50] In other words, electric lighting not only dimmed gas lighting's commercial prospects, it also rendered gas lighting culturally deficient.[51]

Insurers were skeptical of electric light at first, especially since installation protocols were vague and insulation of wires and lamps was often lacking.[52] Electrical workers were unlicensed and, as one observer put it, "do not know the difference between a volt and a pumpkin," while inspectors were often poorly trained. In some cases, insurance premiums for electric lighting were higher than for kerosene and gas.[53] High premiums were explicitly understood to be a hedge against the unknowns of a new technology.[54] As Woodbury said in 1886, electric lighting was "a business without precedents; everything was new, crude and undeveloped."[55] Insurers swiftly set up installation and inspection standards in the 1880s, well in advance of any government regulations, and their concerns then "subsided into a matter of routine inspection."[56] Soon, all parties, from proprietors to foremen to insurers to the workers themselves, agreed that electric lighting was safe and expedient, benefits that resonated powerfully alongside other Progressive Era efforts at industrial reform. Within a short time, new factories were built with electric lighting in mind and old ones were retrofitted to replace gas (figure 4.9).

4.9 (following page)
Enclosed electric arc lamps overhead would cast a bright light throughout, albeit with sharp shadows where blocked by machinery. Lewis Hine, *In the Great Spinning Room—104,000 Spindles—Olympian Cotton Mills, Columbia, S.C.* ca. 1905. Image courtesy of the Underwood and Underwood Glass Stereograph Collection, 1895–1921, Archives Center, National Museum of American History, Smithsonian Institution.

By nearly all accounts, electric lighting was brighter, whiter, steadier, and cleaner than lighting from other sources. It was also closer to daylight in terms of color rendition and the spectral profile for which the human eye had evolved. With no open flame, electric light produced no temperature buildup even after hours of operation. With no combustion occurring, electric lamps consumed no oxygen. Also, electric lamps required far less maintenance than gas lamps. Moreover, they were easier to operate—they could be turned on or off, in whole or in part, at the press of a button. In addition, unlike a gas lamp, the electric lamp could operate in any orientation: "It burns as well upside down as in any other position," noted a contemporary source.[57] Electric lighting also remediated daylight's biggest failing, inconstancy. It remained unbroken throughout the diurnal cycle and across the seasons, and was indifferent to the weather. Woodbury found that owners and managers throughout New England and the Mid-Atlantic states were uniformly satisfied once they had invested in electric lights. His findings moved him to declare that his was the age of greatest advance in lighting.[58] Electricity appeared to have achieved the long-standing ideal of artificial factory lighting: "to light a mill so well that there will be no difference in the character of day and night work."[59]

The Homogenization of Space

Many formerly dark rooms and passageways,
under proper treatment with electric lights,
and by making use of natural light when possible,
have been converted from the liability side to
the asset side of the ledger.
Letter to the editor, *Iron Trade Review*, 1925[60]

In creating visual conditions that were uniform regardless of the time of day or the season, electric lighting also made better use of a mill's spatial resources. With electric lighting, nearly any spot would be amenable to any type of work. Mr. Livermore, an agent for the Manchester, New Hampshire, firm, the Amoskeag Manufacturing Company, perhaps the world's largest cotton textile manufacturer in the early 1880s, suggested that spaces ill-suited for precision work were newly valuable: "I have heard it remarked, and have believed, that weaving-rooms which were dark, by reason of being basements, were now considered as good as those above ground, because the electric light."[61] Similar to the manner in which elevators made upper floors of a building as accessible and therefore as valuable as lower floors, electric lighting made interior spaces uniformly suitable for any type of work. By untethering workspaces from windows, electric lighting allowed factories to be as wide as needed to create efficient arrangements of machinery. An agent for a cotton mill outfitted with Brush arc lamps believed this flexibility in factory floor layout was electric lighting's greatest benefit.[62] In short, electric light's brightness and ease of installation and use unlocked the factory's spatial potential and made it possible to rationalize the floor plan in terms of the logic of production rather than the geometry of access to lighting.

Success with electric lighting opened the door to greater use of electric drives—motors—which revolutionized factory planning more than light could on its own (figure 4.10).[63] At first, electric motors simply replaced steam engines, with the old power transmission system remaining in place. In time they were installed farther down the power distribution system—as group drives running clusters of machines, and then as unit drives, with each machine having a dedicated motor. Ultimately, motors were built directly into machines, allowing them to operate more variably and at greater speeds.[64] Electric drives meant that mills could move closer to markets rather than remain bound to sites with running water, and buildings could be "exactly adapted to contain the required machinery, arranged in the best possible manner to fulfil its functions."[65] Without shaft drives, buildings could

4.10

Individual drives obviated the belt-and-shaft drive system, making interiors brighter and more open. Advertisement, Westinghouse Individual Motor Drive, *Textile World*, July 12, 1919, 21

also be lighter and cheaper to build. Additions would be easier and equipment rearrangement simpler and faster.[66] Moreover, with the ceiling space freed from overhead shafts and belts, cranes could be installed overhead, creating a new work plane over workers' heads for moving materials. Supplies could then flow across the space independent of the configuration of machines below.[67] In sum, the combination of electric lighting and electric drives freed machines and workers alike from the constrained geometries of fixed systems of distributing power with belts and shafts and light through windows.[68]

Henry Ford's Detroit factory in Highland Park, Michigan, was an object demonstration when it opened in 1910 and has remained for historians a landmark in factory design. The plan was large, the spaces were well lit, and the machines were driven entirely by electric motors. Together, electric light and electric drives streamlined factory production as never before and contributed significantly to soaring productivity in the early twentieth century.[69] Contemporaries made the link between the two technologies explicit. Both abolished former limits to the utility of different spaces by making them equally valuable for any productive activity and any sequence of operations; they conferred a consequential elasticity on factory design.[70] In addition, since electric drives could power a single machine, they made it easier to apportion and account for energy use, which facilitated the exact measurement of energy inputs to production, an impossible task with earlier forms of power generation and distribution. To the extent that electric lighting and electric drives were viewed similarly, this accounting efficiency suggested a new threshold for estimating the productive value of judiciously applied electric lighting.

THE DISCOURSE OF PRODUCTIVE VISION

This is the age of "'labor-saving' machinery," but the
greatest of all machines is still the human machine,
and the first condition for its efficient operation is good light.
E. Leavenworth Elliott, 1910[71]

Around the turn of the twentieth century, spurred by ever sharper means of measuring industrial inputs and outputs in the context of increasingly complex manufacturing processes and sophisticated machinery, illuminating engineers and factory managers engaged in a discourse of productive vision. Making use of new photometric tools, they experimented with lighting schemes to meet the detailed visual requirements of factory activities, recommended specialized equipment, developed new accounting tools to better understand lighting costs and explain its advantages, elaborated maintenance programs, and continually sought to demonstrate the benefits of better lighting. Working with the unquestioned dictate of the time that all factors relevant to industrial output could and should be made more efficient, lighting advocates began to ratchet vision into the logical gears of capitalist manufacturing and to assimilate it to a luminously rationalized space of labor.

The Speed of Vision

One's quickness of vision depends upon the degree of illumination supplied. ... There is little likelihood of ever getting our general level of artificial illumination too high.
Ward Harrison, 1920[72]

Light up means speed up.
Electrical Review, 1918[73]

The higher and more variable speeds of manufacturing equipment made possible by electric motors in turn demanded heightened visual alertness from workers. As the historian James Marsden Fitch put it, "A machine is a precision instrument. ... And hence demands visual acuity on the part of its operators."[74] But as gears spun faster, human physiology seemed to have become stuck in another century. Vision was not keeping pace with modernization. In response, lighting engineers tried to calibrate seeing with machine speeds. Woodbury had already suggested in 1905 that a great advantage of electric light was in making possible "the utilization of unavailable time."[75] He was referring to nighttime hours otherwise ill-suited to production, but his phrase also pointed to the idea of time wasted in perception. In the context of the theories of scientific management that were ascendant at the time, outdated ways of seeing objects could be as inefficient as conventional ways of handling them. Concern for efficiency had gone well beyond factory applications and waxed into a desideratum of any organized activity that produced something, whether bolts of fabric or fine moral fiber. Even Progressive Era industrial reform efforts—the political response to the growth of corporate power, new capitalist and management social classes, and greater numbers of people working in factories—generally adhered to the tenets of the "gospel of efficiency."[76] As the polymath critic and historian of technology Lewis Mumford later put it, mechanization of a wide range of activities in the nineteenth century led to a cultural convergence on the criterion of efficiency: "There is only one efficient speed, *faster*."[77] In this context, vision itself, conditioned by a well-lighted space, was ripe for acceleration.

The most ardent exponent of seeing faster was Matthew Luckiesh, an electrical engineer, author, and director of General Electric's NELA Park research laboratory from 1924 to 1950.[78] Luckiesh suggested that seeing had only recently become a new object of study because the act of seeing felt so spontaneous, as if it took no time at all. Luckiesh insisted it did. Seeing was a form of perceptual labor and took time to accomplish. If it took time, it could be sped up; if seeing was not sped up, then it was being wasted. "Of the time consumed in visual discrimination, a portion of it is lost unless the light is best for that particular problem of seeing," he wrote.[79] Thus, Luckiesh argued, when seeing is quickened and perceptual waste eliminated, productivity will rise: "How quickly we can distinguish things determines to a great extent the speed with which we perform the entire act."[80] Framed in this way, lighting was the key to faster and more accurate visual apprehension (figures 4.11, 4.12, and 4.13).

Luckiesh had observed throughout his research subjects squinting, screwing their heads around, and turning objects over in front of them: clear signs of lighting unequal to the visual mission. Having reviewed numerous studies of visual acuity, and after conducting his own ingenious experiments, he concluded that faster seeing required ever higher levels of lighting:

> In general, these data show that speed of vision (of seeing, of recognition, of discrimination) increases as the intensity of illumination increases. By extension it may be safely predicted that, other conditions being equal, a group of workers would turn out more work under a high intensity of illumination than under a relatively low one, when the work involved discrimination of fine details such as in weaving, fine machine work, close inspection, and many other factory operations.[81]

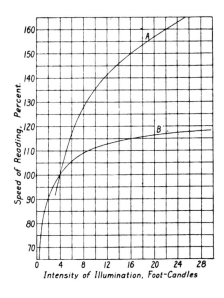

4.11

Under laboratory conditions visual acuity increases with higher levels of lighting, though at a slower rate as intensities increase. Line *A* charts a low-contrast object; line *B* charts a high-contrast object. Subsequent studies would find similar curves obtaining even for much higher levels of illumination. M. Luckiesh et al., "Data Pertaining to Visual Discrimination and Desired Illumination Intensities," *Journal of the Franklin Institute* 192, no. 6 (December 1921): 764.

4.12
Brighter light leads to sharper vision, which leads to faster output and productivity growth. Advertisement, Mazda Lamps, General Electric Co., *Magazine of Business* 47 (1925): 118. Reproduced with permission from the Archives and Research Center, Museum of Innovation and Science, Schenectady, New York.

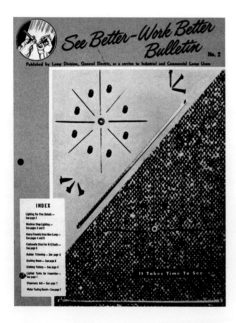

4.13
Seeing takes time. Better seeing equals faster seeing. *See Better–Work Better Bulletin* 1 (Cleveland, OH: General Electric Co., Lamp Division, 1953), cover. Reproduced with permission from the Archives and Research Center, Museum of Innovation and Science, Schenectady, New York.

Luckiesh speculated that lighting levels might need to be even higher in practice since actual workplaces, rather than labs, were rife with inconstant contrasts, uneven backgrounds, unwanted reflections, and other factors that troubled vision. Especially with imperfect conditions, the speed of seeing "continues to increase as the intensity of illumination increases." The upper limit of apprehension, that is, "the maximum speed of vision," seemed to Luckiesh to be a limit set less by optics than by the responsiveness of ocular muscles.[82]

In making the speed of vision amenable to measurement and fine analysis, Luckiesh suggested that workplace suitability might be assessed by the pace of visual perception. He and his colleagues reimagined the efficient body of scientific management proponents such as Frederick Winslow Taylor and Frank and Mary Gilbreth as an efficient eye. When joined to the methods and aims of industrial production, Luckiesh surmised, vision itself might be modernized.[83] This possibility, he pointed out, was grounded in evolution itself. In a logic repeated often by lighting advocates at this time, human eyes developed outside under the high-intensity light of the Sun.[84] Brighter lighting inside therefore not only sped production, it also remedied by technological means a natural condition that had been compromised by modern technological civilization.

The Efficiency of Attention

Our attention is naturally attracted by the objects in a room
which are most brightly illuminated.
J. S. Dow, 1907[85]

The question of attention was very much at stake in research on factory lighting, for Luckiesh and others.[86] After all, seeing faster meant little if workers were unable to concentrate their efforts toward that end. As Jonathan Crary has pointed out, the study of attention flourished in the late nineteenth century, when it was identified as the mental faculty coordinating other mental faculties, charged with selectively synthesizing sensory data into meaningful percepts. Displacing earlier schemes attributing that role to imagination or a generalized "common sense," attention became the means by which the self integrated sense perceptions to form an orderly world view.[87]

Although a mental faculty, attention required a physical environment to deliver sensory impressions, which the self would then organize into perceptual objects. The character of that environment was mostly implicit in philosophical or experimental studies, though the philosopher John Dewey, for one, explained attention by way of analogy to a well-lit thing:

In attention we focus the mind, as the lens takes all the light coming to it, and instead of allowing it to distribute itself evenly concentrates it in a point of great light and heat. So the mind, instead of diffusing consciousness over all the elements presented to it, brings it all to bear upon some one selected point, which stands out with unusual brilliancy and distinctness.[88]

In a sense, Dewey's analogy became a presumption on the part of lighting engineers, though in reverse. That is, for Dewey, attention caused perceptual objects to stand out, as if brightly illuminated. For lighting engineers, a well-lit thing elicited attention. Their discourse of productive lighting posited a material infrastructure of attention characterized by strong lighting proportionate to specific tasks. In such a light-directed environment, attention too could be made more efficient and thereby contribute to greater industrial output.

With attention thus contingent on measurable aspects of the physical context, a subjective mental faculty became an artifact of the objective environment (figures 4.14 and 4.15). Externalized as such, it would be more manipulable and tantamount to "a new political anatomy of the body," as Foucault described the

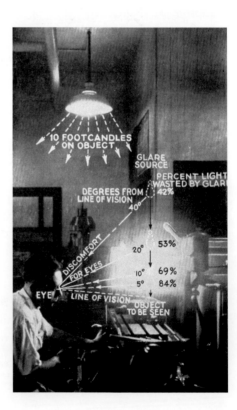

4.14

An environment reconfigured by the geometry of lighting. Vectors emanating from the worker's eyes replicate vectors radiating from the lamp to form a portrait of exteriorized attention. Matthew Luckiesh and Frank K. Moss, *Lighting for Seeing* (Cleveland, OH: General Electric Co., NELA Park Engineering Dept., January 1931), 27. Reproduced with permission from the Archives and Research Center, Museum of Innovation and Science, Schenectady, New York.

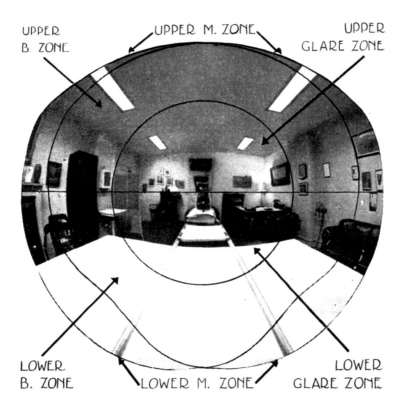

4.15

Unlike most other depictions of lighting conditions in the workplace, this study attempts to account for the subjective experience of lighting for, in this case, an executive. Captioned "Field of view of an executive." H. L. Logan, "Light for Living," reprint of paper delivered at the annual convention of the IES, Quebec, September 18–20, 1946, p. 1. Subsequently printed in *Illuminating Engineering* 42 (March 1947): 291–324.

implementation of any technology of power aimed at the restructuring of human subjects, the formation of "calculable man," in Foucault's terminology.[89] Concerns regarding the luminously rationalized space led in this way to ideas of the luminously rationalized subject who, in internalizing such visual conditions, becomes an instrument of self-regulation. Just as early nineteenth-century managers used clocks to synchronize the internal time sense of laborers with the rhythms of factory work, so did illuminating engineers conceive of lighting as a means to coordinate seeing with industrial production. To the degree that workers in turn demanded good lighting, they became personally accountable for the expedient deployment of their own vision. In this way, uniform ambient electric light peppered with bright spots for close looking became both a medium for and an image of productive attention.

Illuminating the Balance Sheet

Good lighting increases production.
William Durgin, 1918[90]

However much illuminating engineers maximized lighting's contribution to production, it had still to be legitimated in hard numbers. Electric lighting was an expense and mill owners had to capitalize it. They had to ask themselves, how much was electric light worth?[91] With advances in photometry, measuring light was becoming straightforward; measuring gains in an individual's visual acuity was possible, at least in controlled settings; but measuring the productive gains attributable to greater acuity was far more difficult. In the 1890s and early 1900s, the cost of operating electric lighting was often compared to the cost of using gas (less, in some accounts) or kerosene (usually more), as well as to daylight (substantially more, with important qualifications regarding daylight's variability), followed by assertions about electric light's other benefits.

Early advocates such as Woodbury presented extra-economic arguments, declaring that electric lighting's cost "bears little relation to its intrinsic worth."[92] Some acknowledged that any direct savings were offset by hidden costs, such as wages for mechanics to run the dynamos. Its value lay elsewhere, Mr. Livermore, the agent for Amoskeag, stressed: "The saving will not appear in the balance-sheet as to light." Rather, it showed up in the quality and quantity of output, with no seasonal differences in either, a result he had seen at his own mills.[93] Foremen and managers in the late nineteenth century often offered similar anecdotes, but with little evidence. At the time there were no methods with which to monetize electric lighting's "collateral advantages," in Woodbury's phrase, such as steadiness, brightness, flexibility, and no flame to vitiate air or raise room temperatures.[94] The bluster of "electro-mania" muddied the problem further, complained an insurance agent, beleaguered by extravagant claims made by patentees seeking approval from underwriters.[95] Such vague gestures toward value were increasingly deemed inadequate to justify the mounting costs of electrification, the growing scale of industrial production, and rising standards in accounting practices. Light needed a more reliable price tag.

Charles Felton Scott, a professor of electrical engineering at Yale University's Sheffield Scientific School who had worked with the visionary physicist and electrical engineer Nikola Tesla, took a major theoretical step forward with a 1910 editorial in *Electric Journal*, a trade publication he had founded.[96] Although hypothetical in detail, his logic was clear: assess the cost of lighting in terms of minutes of labor saved thanks to better vision. The cost of light should be weighed alongside all factors of production, he said, which could be made comparable in like fashion. Assessing electric lighting on this basis, Scott arrived at the claim that the cost of a good lighting system over a bad one was "about five minutes" of a working day. If, he concluded, improved lighting meant that "more and better work is done in eight hours than would have been done in eight hours and five minutes with the poor light, then the extra cost is justified." He frankly acknowledged that more precise measurements would be vexing but that it was likely that "better light and better vision" would result in productivity gains of up to 50 percent. Even a 1 percent gain would return three times the investment. Initial installation costs, Scott insisted, were insignificant in relation to costs over time to operate the system and, on the positive side, projected gains in productivity. It was a premonition of what would come in accounting circles to be called "life-cycle costing." Precise calculations were complicated, he conceded, and necessarily included nebulous factors, which, he stressed, constituted a great deal of electric lighting's value.[97] While conjectural, Scott's expression of the cost of electric lighting as a percentage of wages, expressed in minutes, sharpened discussions about the value of light and made the idea of enhanced vision, and the luminous spaces that fostered it, more amenable to rational economic analysis.

Regardless of the conjectural bridges Scott constructed, his evaluative structure proved to be persuasive and was soon cited as an objective method of analysis and, importantly, a way to sway hesitant factory owners. An editorial in the *American Architect* published just months after Scott's article invoked a factory's "curve of efficiency" in a scientific-sounding claim that lighting had finally obtained a "rightful role as an important factor in calculating the cost of production." Factory owners ought to understand that they were not just buying electrical current; they were buying light and all its attendant benefits, the editorial emphasized. Electrical

4.16

Production leaps with higher and higher levels of lighting. Clewell asserted that good lighting was crucial to production but its costs were negligible in relation to expenses for wages, interest, depreciation, materials, and the like C. E. Clewell, "Economic Aspects of Industrial Lighting," *Electrical World* 73, no. 8 (Feb. 22, 1919): 371.

4.17

Lighting as an incentive to productivity growth. Similar advertisements in this campaign encouraged managers to think of lighting as an investment rather than an expense. Advertisement, Westinghouse Lamp Co., *Textile World* 59, no. 6 (Feb. 5, 1921): 128.

current was an expenditure; light was a resource that helped a plant extract higher profits from raw materials: "Adequate illumination is thus seen to be one of the few basic essentials to high manufacturing efficiency."[98] By likening electric light to other manufacturing costs, Scott launched a new stage in the assimilation of vision to productive purposes.

Clarence Edward Clewell was Scott's colleague in electrical engineering at the Sheffield School. He was hired in 1912, a year after Scott, and moved to the University of Pennsylvania in 1914. Active in the Illuminating Engineering Society (IES), he chaired a committee charged with drafting an industrial lighting code and wrote numerous articles on the topic.[99] In his 1913 booklet, *Factory Lighting*, he described Scott's discussion of "artificial lighting in terms of wages" as a breakthrough.[100] He was writing, he noted, to flesh out Scott's insights with data gathered from case studies (figure 4.16). He concentrated on economic questions about the quality of workspaces, he said, because that was what management did.[101] With recent developments in lighting, including a range of new lamp types, factory owners needed new evaluative tools applicable across a range of conditions.[102] In all cases, to make light fungible in accounting terms, it had to be translated into the common currency of capital investment valuation.[103] His task was made easier, Clewell noted, by a nationwide "change in attitude" regarding factory design, which included a wave of public support for recent efforts to regulate the visual environment of workplaces.[104] He thought legislation would be unnecessary if managers could only be persuaded to view lighting simply "as an asset to the output of the plant" rather than worrying over capital costs or ongoing operating costs (figure 4.17).[105]

With World War I placing greater demand on American industry, Clewell and other electrical engineers were quick to note that better lighting was the fastest way to increase production. William Durgin, who had earlier supervised a comprehensive industrial lighting study for Commonwealth Edison in Chicago, repurposed the idea of "productive intensities" to denote its original meaning, the relative degree to which a single input affected production, and the actual brightness of the lighting. In doing so he further cemented higher factory output to higher levels of lighting.[106] The phrase was immediately deemed "a very happy expression" that would be especially meaningful to management.[107] It was also a tool for the patriotic engineer "to apply light to war."[108] Clewell himself, citing Durgin and others, subsequently asserted that benefits of better lighting accrued not only to the factory owner but to the nation as a whole, with energy savings redirected to the war effort.[109] After the end of hostilities, lighting advocates stressed that even without wartime incentives, high levels of lighting would lower labor costs for production, referring back to Durgin's studies.[110]

Establishing common methods of cost accounting is one step toward establishing shared and durable conventions applicable to various activities or things. Such standards create a kind of order from heterogeneous actions and objects (figure 4.18).[111] They replace individual values and abilities with institutionally desirable skills that advance a set of goals.[112] While quantification is prerequisite to the process, standardization nonetheless embodies social values, which are typically the basis for evaluating them. That is, the legitimation of a standard exceeds the subject of that standard, whether by appeal to categories of efficiency, public safety, community well-being, or, as happens in wartime, patriotism. In establishing a standard, it is essential to identify a market for it, that is, to define a community that materially benefits from having accepted practices. The broader that community, the more influential the standard.[113]

4.18
Factory managers were encouraged to replace their subjective judgment of lighting with objective standards in order to reap the productivity gains that came with brighter lighting. Advertisement, Mazda Lamps, General Electric Co., *Saturday Evening Post*, Feb. 14, 1920.

Insurance underwriters were the first to establish standards for the electrical industry, after restricting policies following mill fires in the early 1880s. Within a few years, they began collaborating on a code focused on power generation, equipment manufacturing, wiring, installation, and safety provisions.[114] They improved it over the years, earning endorsement for it as "truly an American standard." Public safety in electrification advanced hand in hand with predictable insurance markets.[115] In the early twentieth century, states developed lighting codes within more general labor laws. In 1913, New York was one of the first states to consult directly with the IES. The IES issued its own Code of Lighting for Factories, Mills and Other Work Places in 1915, based to a large extent on Wisconsin's 1912 regulations, after consulting legislative committees, public service commissions, manufacturers, insurance companies, installation experts, and school boards. Recognizing that every lighting installation was unique, the IES code featured "representative cases." Although workplace conditions varied, there were several main objectives: fewer accidents, more accurate workmanship, greater productivity, less eye strain, better working conditions, "greater contentment of the workmen," "more order and neatness in the plant," and "supervision of the men made easier."[116] The code offered formal methods for achieving these heterogeneous goals within the same analytical frame, "visual efficiency." Since nearly all activities informed by vision could be similarly analyzed, all of society, it was argued, would benefit from standards for better seeing.[117]

THE MORALITY OF BETTER SEEING

Industrial waste figures cannot compare to waste from bad lighting.
E. Leavenworth Elliott, 1910[118]

Good lighting is a foe to disorderly arrangement and uncleanliness because it exposes these slovenly conditions about a plant and, by making the employes more alert, it makes them ashamed of contributing to disorder and filthiness.
F. H. Bernhard, 1918[119]

Although many engineers subscribed to a technocratic view of labor, worker safety and sense of well-being were nearly always paramount in any study of factory lighting, especially during the Progressive Era. Matthew Luckiesh, who compared the eye to an automobile transmission and an electric motor, took care to assert that good lighting was not coercive. Laborers were not forced to work faster or harder; they simply wanted to because good lighting made work more satisfying: "By increasing the ease or certainty of seeing it makes work easier and acts more

4.19

Reconceiving the workforce as an assembly of eyes warranting management and optimization. W. C. Brown and Dean M. Warren, *Lighting for Seeing in the Office* (NELA Park, OH: General Electric Co., NELA Park Engineering Dept., 1936), 4. Reproduced with permission from the Archives and Research Center, Museum of Innovation and Science, Schenectady, New York.

4.20

The morality of better lighting is made clear in a series of advertisements depicting workers blinded, fatigued, and fearful. Advertisement, Benjamin Industrial Lighting, *Factory* 22, no. 6 (June, 1919): 1353.

certain. ... The worker is working just as hard (and with greater annoyance) under poor lighting as under good lighting but he is not accomplishing as much."[120] At some point, nearly all those concerned with workplace lighting advanced similar reasons from the workers' point of view, along with claims of improved production. What developed was a morality of the well-lit workplace, which stood on several legs: it was better for workers, better for management and corporate identity, and better for American society as a whole (figure 4.19).

Worker safety routinely topped the list of benefits conferred by better lighting. Accidents in factories were threats to individual workers, costly for management, damaging to the corporate brand, and bad for workforce morale. Study after study, not to mention common sense, demonstrated that good lighting resulted in fewer accidents. The relationship of lighting to worker safety became a matter of public health. A 1915 report from the *British Medical Journal*, to take only one example, noting the nighttime and overtime hours necessary to meet wartime government orders, lauded lighting levels that increased both productivity and safety.[121]

Workers' eyes in particular had been a concern since the nineteenth century, especially in trades demanding precise and detailed work. Numerous sources attest to the visual problems of laboring by subdued light. Even the brighter light of gas lamps was no help, wrote Friedrich Engels, who found it "very bad for the eyes, ... inflammations of the eye, pain, tears, and momentary uncertainty of vision during the act of threading are engendered. For the winders, ... frequent inflammations of the cornea, many cases of amaurosis and cataract."[122] Labor-oriented publications agreed that even with gaslight, mill workers suffered "weakness of eyes, from the performance of difficult work by lamp-light."[123] In addition to the specific diseases critics mentioned, a more general disorder of "visual fatigue" began to be cited for its effect on workers (figure 4.20).

Fatigue, as a category of malaise, derived from nineteenth-century theories of energy, most famously in the physiologist, physician, and physicist Hermann von Helmholtz's formulation of the conservation of energy in a fixed system. Said to be valid for physiological as well as mechanical energy, it provided a resonant scientific rhetoric for relating the industrial order to the natural order. The argument at the time went as follows: Like machines, human bodies needed energy to operate; some of that energy would advance an objective, while some of it would be wasted; wasted energy was not just lost, it also generated unwanted by-products that could be toxic if not expelled. With this powerful abstraction, all physical and social phenomena could be framed as operating within common cycles of energy acquisition, expenditure, and waste.[124] Electricity as one such form of energy was subject to the same laws of energy conservation, whether running through nerve cells or telegraph wires or causing filaments to glow.[125] In this context, workers were mechanisms that expended energy in a more or less efficient manner. Likewise, component activities, such as seeing, concentrating, and other types of "mental labor," could be analyzed in terms of an efficient use of energy. In this way, workplace activities could be normalized and addressed as more or less organic manifestations of natural processes.[126]

While the contribution of social factors to workers' fatigue was noted by such figures as Marx, proponents of scientific management exploited a mechanistic understanding of fatigue to advance the cause of industrial efficiency in particular.[127] As it was generally understood, "fatigue is the antithesis to efficiency," according to a lead investigator with the British government's Industrial Fatigue Research Board, established in 1918.[128] With the impetus of bourgeoning fatigue studies in the early twentieth century, as well as Progressive Era industrial reform efforts, visual fatigue came increasingly under scrutiny.[129] Although Frederick Winslow Taylor did not discuss lighting in any sustained fashion, Frank Gilbreth, one-time exponent of Taylorism and concerned perhaps more with worker well-being than with industrial output, did. Lighting, he argued, was too often designed

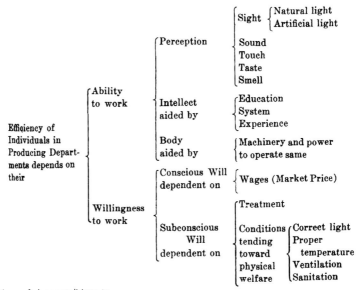

4.21
The pyramid of relations of shop conditions to worker efficiency: lighting joins other factors that bear on the body, the conscious and unconscious mind, and perception to affect workers' ability and will, both of which underpin productivity. F. B. Allen, "Important Considerations in Factory Lighting," *Electrical Review and Western Electrician* 59, no. 7 (Aug. 12, 1911): 323.

to meet the convenience of the contractor installing and the manager overseeing a system rather than the visual needs of workers who labored by it day after day. The result was plain to see: "Go into any factory and examine every light, and you'll notice that as a rule they are obviously wrong."[130] Never put the comptroller in charge of the lighting, he urged. Lighting "is the cheapest thing there is," and cutting corners on it was a false economy once the question of fatigue was taken into account. "The difference between the cost of the best lighting and the poorest is nothing compared with the saving in money due to decreased time for rest period due to less fatigued eyes." He stressed that the worker's whole body must be comfortable, including the eyes, a principle ignored, he observed, in most workplaces.[131] Good lighting was the primary defense against the threat of visual fatigue (figure 4.21).

More than simply depleting the body's store of energy required for physical and mental labor, visual fatigue depressed the very will to work, researchers claimed.[132] Electric lighting, again, was revitalizing, more or less in the same way that sunlight was. Plants could grow under it, providing "every reason to believe that the electric light, which is so valuable a stimulant to vegetable energy, has an equally stimulating effect on the animal energy."[133] Edward Atkinson, a cotton entrepreneur, reported in 1883 hearing from mill foremen that under electric lighting, production improved, "owing to the more vigorous condition and vitality of the operatives" attending the machines.[134] Psychologists as well reported on the energizing effect of well-lit spaces: "The effect is like a stimulant," according to a 1903 report, with test subjects feeling "full of life and nervous energy" and able to think more quickly.[135] Light's "stimulating effect" was a recurrent trope in the early twentieth-century literature on factory lighting, with light bulbs regularly likened to the sun itself.[136] Electric lighting initiated what seemed to many to be a self-evident chain of symmetries: its effects were like those of natural light; its rousing visual atmosphere aroused mental activity; brightened eyes were a sign of rapt concentration; and a sharp attention could operate even the fastest machinery.

The strongest element of the moral argument for electric lighting was that workers themselves clamored for it. One mill owner testified in 1882 that his employees asked to be reassigned to those of his buildings that had recently been fitted with electric lights.[137] Others explained that workers were in the vanguard of lobbying for better lighting. Trade journals carried reports of agitation in Europe in favor of higher lighting intensities. The authors of a 1909 account described employees buying their own lamps after managers at a textile mill in Pennsylvania removed tungsten bulbs following an experiment assessing the color rendition of different light sources, workers in New Jersey buying higher wattage bulbs with their own wages and masking the resulting glare with makeshift paper bag shades, and employment applications surging at a mill in the South after it put in electric lights. Where factory owners failed to see electric light's advantages, workers understood immediately, the authors asserted.[138] Here was one concession to labor that management was widely encouraged to make on moral grounds.

By the early twentieth century, engineers usually called for a distribution plan that combined ambient lighting with localized task lighting for most factory settings. The resulting visual field was more uniform than in the past, with diverse activities scattered around the factory floor looking like so many elements of a common undertaking (figures 4.22 and 4.23). In many discussions, the "cheerful appearance" generated by ambient lighting stimulated "the idea of wideawakeness"

Too wide spacing results in poor distribution, as shown by the top illustration. For best seeing conditions, the illumination throughout the work area should be reasonably uniform. By properly spacing the outlets, as illustrated below, uniformity results.

4.22

Light-mapping exercise to visually argue that uniform lighting schemes offer the "best seeing conditions" in the workplace. Allen Kiefer Gaetjens and Dean M. Warren, *Lighting for Production in the Factory* (Cleveland, OH: General Electric Co., NELA Park Engineering Dept., January 1939), 7. Reproduced with permission from the Archives and Research Center, Museum of Innovation and Science, Schenectady, New York.

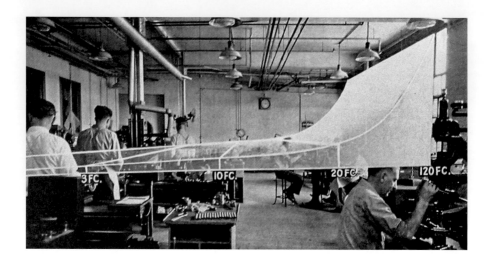

UNSAFE, UNPRODUCTIVE LIGHTING WORTHY OF THE DARK AGES

THE SAME FACTORY MADE SAFE, CHEERFUL, AND MORE PRODUCTIVE BY MODERN LIGHTING

4.23

Light mapping shows the rapid decline of lighting with distance from a window to underpin a sense of visual injustice that would be remedied with electric light. Allen Kiefer Gaetjens and Dean M. Warren, *Lighting for Production in the Factory* (Cleveland, OH: General Electric Co., NELA Park Engineering Dept., Co., January 1939), 24. Reproduced with permission from the Archives and Research Center, Museum of Innovation and Science, Schenectady, New York.

4.24

Before and after: from medieval deprivation to modern amenity with new factory lighting. Matthew Luckiesh, *Artificial Light: Its Influence upon Civilization* (New York: Century Co., 1920), opp. p. 233.

and presented the mill "as a whole, well lighted, giving a cheerful and wide-awake appearance at night."[139] Uniform lighting hinted that all workers were treated equally, and the open field of view it established facilitated supervision and, from the workers' perspective, awareness of supervision. In quite a few discussions, however, supervision would not be necessary as workers came to internalize best practices under electric lighting. As Clewell wrote in 1918, workers were self-policing in "cheerful surroundings … with little or no verging on the part of foreman and superintendents."[140] In this way, the beneficence of the luminously rationalized workplace permeated workers' spirits to make their individual aims one with the corporate mission. It created an optical imaginary wherein actual differences in work activities would be diminished in favor of a mutual visual experience. Seeing in the factory would be as much about the visual conditions binding worker to machine as the ability to see one's work well. Like grease in a machine's bearings, good lighting lubricated seeing and in so doing burnished corporate identity with an image of progress: the well-lit space of production (figure 4.24).

Exemplifying Welsh cultural critic Raymond Williams's definition, the well-lit factory embodied a "whole live social process" organized by institutional aims and joined voluntarily by its constituents in a luminous portrait of corporate hegemony. As such, it was built on normative perceptions and practices that were self-validating, forming a fabric of experiential reality whose constructed nature was only available analytically.[141] Hierarchical relations in the factory would be veiled when a key reference point for worker productivity was a visual norm rather than an overt exercise of power. Moreover, such descriptions of workers' enthusiasm for electric lighting preemptively eliminated the chance for counterhegemonic actions, such as workers' intentional production slowdowns described by E. P. Thompson. In Thompson's account, clocks in the first factories instilled in workers a uniform sense of time that gradually displaced workers' often idiosyncratic, pre-industrial time sense. Similarly, electric lighting set new norms that reduced irregularities in production and indiscipline in the workforce.[142] The moral underpinnings of the discourse of productive vision served in this way to naturalize the visual conditions of labor and to standardize sensory experience within the workplace.

THE FRANCHISE FOR BETTER SEEING

The importance of good lighting in relation to the eyesight of school-children and of adults in factories has often been emphasised, but the broader question of its effect on the eyesight of the nation is one that should not be overlooked.

Leon Gaster, 1918[143]

As arguments for the productive gains promoted by improved lighting piled up, other spaces of work came to be scrutinized for their visual conditions. Schools, shops, offices, even the home: all were places where something was produced, whether knowledge and citizenship, sales or services, or family life and rest. All could benefit from precisely targeted lighting. Shops, for example, could be luminously recalibrated to enhance the appearance of their wares. Indeed, the trade literature on retail lighting offers a sustained study of the consumer mind in relation to visual perception. Homes too were dissected for the various sorts of illuminated work that took place within them. Cooking, cleaning, reading, playing games, or putting on makeup: all were goal-oriented activities that could be made more efficient with appropriate lighting.[144]

Bad-Eye Factories

The strain of modern civilized life falls most heavily
upon the eyes, the hardest worked and
the most delicate of all organs of the body.
Reginald Augustine, 1922[145]

More than any other building type, schools were likened to factories. Although forms of public education varied, models such as the nineteenth-century Monitorial System or the Lancaster System resonated with contemporaneous ideas regarding factory work, including the assembly of groups of individuals in large rooms to perform standardized operations in an effort to achieve efficiencies of scale. As the pioneer educational administrator Ellwood P. Cubberley put it in 1916: "Our schools are, in a sense, factories, in which the raw products (children) are to be shaped and fashioned into products to meet the various demands of life."[146] Vision was a key element of that process. Poor lighting made both teaching and learning harder and was a particular hindrance to "the cultivation of the faculty of attention," which required readily discernible objects for students to concentrate on.[147] Inadequately lighted schools were sometimes described as "bad-eye factories," a phrase attributed to a Dr. Scripture at Yale University in the 1890s.[148]

The widely held argument from evolution, noted earlier in this chapter, underpinned the analogy between schools and factories: our eyes evolved for distant vision in life outdoors; modern life was conducted mostly indoors and entailed "near work" and "eye-work," terms used across a number of industries to describe machine operation, handicrafts, and, importantly, reading and writing (figure 4.25). Such occupations demand a "minute correlation of physiological functions to environmental conditions," that is, close calibration of visual needs with lighting conditions, regardless of the exact nature of the work involved.[149] Factories were in the vanguard in this regard, as Reginald Augustine, a former president of the American Optometric Association, pointed out in 1922, and schools, he continued, should look to them for cues to better lighting to make education more efficient.[150] Like fatigue in factory work, students seeing under less-than-optimal conditions in schools was pathologized.[151] Bad lighting led to a range of maladies, with myopia chief among the optical consequences. Studies were said to show that myopia increased steadily in children after they started school because of bad lighting.[152] Augustine cited findings that only 10 to 12 percent of children had "normal eyes," leaving all others susceptible to eye strain from poor lighting (figure 4.26).[153]

Bad lighting also deformed spines (figure 4.27). Authorities argued that vision and posture were closely related: students instinctively contorted themselves to minimize visual discomfort. As students attempted to realign their bodies to ambient visual conditions, good lighting could serve as an environmental prosthetic, able to straighten the spines of students as they practiced their lessons.

◄ The eyes evolved for distance vision under thousands of footcandles. The illustration shows a common seeing task.

Today's work, for the most part, is done indoors and at close range. Seeing tasks are difficult and in some instances prolonged.
▼

4.25
The argument from nature: human eyes evolved for life outdoors but modern labor is conducted indoors. *Lighting for Seeing in the Office* (Cleveland, OH: General Electric Co., NELA Park Engineering Dept., December 1936), 5. Reproduced with permission from the Archives and Research Center, Museum of Innovation and Science, Schenectady, New York.

4.26
Measuring the optimal distance from desk to eye under specific lighting conditions, based on prior experience in factory work. Stuart H. Rowe, *The Lighting of School-Rooms* (London: Longmans, Green, 1904), opp. p. 68.

 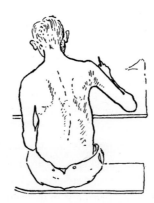

4.27
Bad lighting may become embodied as children contort themselves to see better and unwittingly injure their spines. Leon Gaster, "The Economic and Hygienic Value of Good Illumination," *Journal of the Royal Society of Arts* 61, no. 3142 (Feb. 7, 1915): 305.

Light, long a metaphor for knowledge in mottos such as "Lux et Veritas," was made physical in these discussions, not simply an image of learning but a factor in the making of sound bodies along with sound minds. In response, school boards were encouraged by groups such as the American Posture League or agencies such as the Bureau of Education to assess conditions and improve school lighting.[154] At the same time, many states developed lighting codes for schools, often based on earlier codes for factory lighting.[155]

Better Light–Better Sight

"When we found out that Jack needed glasses we also learned to our dismay that he had *never* had enough light to study by."
USDA Housekeepers' Chat Radio Service, 1937[156]

The Better Light–Better Sight campaign, launched in 1933 under the auspices of the Edison Electrical Institute, was the industry-wide campaign to appeal directly to the largest constituency of all: consumers. It was endorsed by those who sold or installed not just electrical equipment but virtually anything that reflected light, including the National Paint, Varnish and Lacquer Association, the National Retail Furniture Association, and the National Retailers Contractors Association, not to mention the Illuminating Glassware Guild, the Illuminating Engineering Society, and the National Electric Light Association. Active for decades, the National Better Light–Better Sight Bureau probably did more than any other group to spread the gospel of better lighting to all venues. In most cases this meant more lighting. The bureau published materials under its own imprint, the Better Light–Better Sight Bureau Press, and produced films for general viewing, such as the 1936 *Owed to Your Eyes*. It sponsored radio programs that told of anguished mothers learning that their children needed eyeglasses because they had failed to provide enough lighting at home, a shortcoming resolved with more lamps, brighter bulbs, and more electrical outlets throughout the house.[157] It also set industry campaign themes and endorsed products such as light meters for home use and Better Light–Better Sight Lamps, which carried special tags to assure buyers. And it drafted text incorporated by the U.S. government in its public education efforts.[158]

The speed of vision and lighting's enhancement of productivity were consistent themes of Better Light–Better Sight drives. The campaign owed a clear debt to Luckiesh, who just a few years earlier had proposed, along with the physicist Frank Moss, a "new science of seeing" that supplemented engineers' technical methods of light measurement for individual tasks with a holistic approach to visual perception more generally. From this perspective, Luckiesh and Moss identified optimal lighting conditions for easier, more accurate, safer, faster, and more comfortable seeing, all of which converged on matters of productivity and the conservation of vision, that is, seeing without wasted mental or physical effort.[159] The reception of their work varied. One review of a later version of their writings acknowledged the novel experiments but held it was less a compilation of disinterested empirical research than a misleadingly titled document of corporate "evangelism" that had nevertheless helped to establish ever higher levels of lighting.[160]

Along with many corporations, General Electric recognized the importance of appealing directly to consumers as a way to boost revenues. Their "light conditioning" campaign of the early 1950s scaled illuminating engineers' techniques, such as light metering and luminous mapping, to the different rooms and activities of the home. Drawing on themes in contemporaneous suburban culture, the campaign emphasized easy-to-follow "tested lighting recipes" and stressed that a home could not be genuinely modern unless it conformed to modern standards of lighting (figure 4.28).[161] General Electric had floated the theme earlier as a means of bringing "better working conditions and new joy in living to young and old alike in millions of homes, stores, offices, factories and schools." The industry claimed to employ thousands of "light conditioning experts" available for consultation. Homeowners were encouraged to invite a "home lighting advisor" to bring light meters and branded bulbs. The advisor would measure light levels wherever "seeing tasks are done." A major claim of the legitimacy of the effort was that "carefully conducted experiments" in factories and schools had increased production and improved grades, respectively.[162]

4.28

"Light conditioning," a sustained promotional campaign introduced by General Electric, linked the physiology of vision with management paraphernalia of everyday domestic life. Advertisement, General Electric Co., *Life Magazine*, Sept. 24, 1951, 128. Reproduced with permission from the Archives and Research Center, Museum of Innovation and Science, Schenectady, New York.

Such campaigns were enormously successful. A wartime report noted that commercial lighting intensities had "increased markedly during the last ten years" as a result of these campaigns and it anticipated even higher intensities after the war thanks to promotions aimed at "restimulating customers to increase intensities of illumination to adequate levels for each particular 'seeing' job."[163] Promotions were so effective that recommended and actual lighting levels for a range of activities increased steadily for decades. Office lighting recommendations by the General Services Administration, for example, went from 15 foot-candles in the 1930s to 75 foot-candles by the 1960s.[164] Recommended levels for "general interior" illumination rose from 3 foot-candles in 1900 to 125 foot-candles in 1971. To achieve these levels, businesses and property owners had to buy vastly more equipment and electrical current.[165]

In the 1970s, social reformers such as Ralph Nader highlighted collusion in the lighting industry and promotional campaigns that smuggled sales pitches for more equipment and more current under the guise of improving public health. Soon, claims regarding the need for ever higher lighting levels, as well as efforts to pathologize visual activities such as reading in relatively dim light, were being refuted, especially after the 1973 energy crisis.[166] A 1974 report, for example, pointed to the conflict of interest between manufacturers of lighting equipment and members of the IES, many of whom received fees as corporate consultants or worked directly for power and lighting companies. The report charged that, with nearly one quarter of the organization's income derived directly from manufacturers, retailers, and central stations, some recommended standards for higher lighting intensities above certain thresholds were suspect.[167]

CONCLUSION

The tax returns from which employees work are
illuminated for easy, comfortable seeing.
Result: faster, more accurate work from happier workers.
See Better–Work Better Bulletin, 1959[168]

With the Better Light–Better Sight and Light Conditioning campaigns, along with numerous other professional, trade, and legislative efforts, an industrial rationale regarding vision was broadly assimilated in cultural terms to the extent that a bright, well-lighted place became both a vehicle for greater productivity and the image of an efficient enterprise. The long history of lighting for mills and factories proved a prelude to visual rationalization practices that found their way into other social spaces as a set of optimal optical rules relevant to labor, however loosely labor was construed. Places of work came to be understood as spaces of visual effort, with constructive seeing a central element of industrious activity.

Along with the brighter lighting standards established over the decades by industrial illuminating engineers, a visual profile of the industrious laborer was also produced and subsequently adopted as a figure of the effective citizen. The luminously rationalized space of labor had generated a rational seeing subject, whose vision was historically formulated by lighting professionals aiming to meet the demands of industrial production. This model of vision was as much a part of its spatial context as Foucault's model of surveillance was based on Jeremy Bentham's design for a Panopticon prison, as much as Charles Baudelaire's depiction of the flaneur's gaze was prompted by Parisian boulevards, and as much as Siegfried Kracauer's discussion of distracted seeing was born of cinema and other collective displays. Notwithstanding some physiologically necessary minimum lighting levels for effective seeing, the naturalization of brightness norms in relation to types of work and the cultural diffusion of visual standards to a range of settings was animated by the fact that electric lighting not only facilitated visual attention, it also represented attention. A well-lighted place became as much a figure of sharpened consciousness as it was the spatial embodiment of modern seeing.

5
ELECTRIC SPEECH IN THE CITY

INTRODUCTION

Electric-light advertising is going to do big things if it is given the least chance.
SMILAX, 1905[1]

A new urban district comprised of electric light appeared in the twentieth century: the zone of illuminated commercial speech. It emerged most decisively at Times Square, New York City's newly formed theater and entertainment district (figure 5.1). It was created with "sky signs" or "spectaculars," large, façade- or rooftop-mounted metal frameworks supporting lighted advertisements. They appeared sporadically in other cities, but they flourished in Times Square at the start of the twentieth century. There, as Broadway sliced through Manhattan's street grid to form a dynamic, diagonal volume, outdoor advertisers put up signs that created shifting scenes acted out by spirited figures and leaping letters. In doing so, they invented an urban space that was an entertainment in itself. People began to come to Times Square as much for the signs as for the restaurants and theaters. They came to gaze skyward and see letters line up to spell words, then flicker, flash, and dissolve back into the night, only to repeat the cycle again. Circling around the square, they looked from left to right as one sign after another, vying for attention, sparkled its advertising message at those standing below. In this new type of urban space, visitors to Times Square also learned a new way to read in the city.

For millions, it was also a theater of American commerce, its jumpy energy the epitome of high-spirited and unfettered competition. Urban districts had earlier been dedicated to both commerce and entertainment, but none had ever coalesced into a definite urban figure, or "civic presence," as the historian William R. Taylor put it, until the industrial age, with Times Square unique in its use of electric light and in its luminous colonization of public space by private interests.[2] Describing Times Square's "superabundance of signs" and its sense of "overfullness," the historian and cultural critic Marshall Berman wrote: "The development of Cubism in the early twentieth century was made for spaces like this, where we occupy many different points of view while standing nearly still. Times Square is a place where Cubism is realism."[3] The elements of Times Square's lived Cubism, its "special contribution to modernism," he continued, composed "a spectacular trinity—people, lights, sky—as the new totality of being."[4] In collapsing onlookers, business interests, a new electric technology, and the urban setting into a single, dazzling space, Times Square also became a landmark of modernity, a permanent fixture of New York life built on frothy evanescence. It was a new urban destination, founded on flickering electric light, forged in the canyons of Manhattan and subsequently spun off to cities around the world like sparks from a flame.

5.1
Lighted commercial speech in Times Square, ca. 1923. The J. Paul Getty Museum, Los Angeles. Unknown maker, American. New York Edison Co. Photographic Bureau, 1923, gelatin silver print, 17.8 × 22.8 cm.

MAKING SIGNS

The primary purpose of the large electric sign of the Broadway
type is to send its message on a national scale rather
than to try to influence the individual to stop at once and
buy a new suit of underwear.
The Edison Monthly, 1920[5]

Commercial speech had long been an element of city life in the form of shop signs, trade cards, handbills and broadsides, newspapers, banners, posters, billboards, shop window signs, and theater marquees. Nineteenth-century American cities suffered a plague of bill posting that blemished the buildings on which the bills were posted like a skin disease. New printing techniques only broadened the franchise for the form and extent of urban text. As a French visitor to New York noted in 1864, "Advertising is the indispensable adjunct of this great village fair. On every hand are floating banners, monstrous signs, flamboyant decoration. Advertising matter extends into the street, onto the edge of the sidewalk between the gutter and the pedestrians' feet."[6] A crust of advertising and printed matter settled across the city, becoming a constituent element of urban public space.[7] Likewise, lighted text had long been part of urban life. After sundown, candles, oil lamps, and eventually gas lamps brightened shop windows and kept signs legible in the dark. Savvy bill posters pasted their sheets beneath gas street lamps to remain effective at night at the city's expense. With electric bulbs, signs themselves could radiate light and not just reflect it.

Lighted billboards appeared in Paris and London in the late nineteenth century and attracted a great deal of attention. But they were not as concentrated as those in Times Square would prove to be, and they also set off a backlash that resulted in a series of prohibitions that slowed their spread.[8] In New York City, however, lighted signs were an integral part of the growth and spread of the city's theater district. According to the eminent theater historian Mary Henderson, New York invented the theater district as a distinct "architectural entity," predicated on the formation of a permanent audience eager for nighttime diversions such as seeing plays and dining out.[9] As the theater district moved north over the course of the nineteenth century, from lower Broadway to Union Square then Madison Square then Herald Square, animated lights were part of its character. Theaters and shops similarly used oil lamps and candles to light interiors and to make marquees and signs visible by night. Gas street lighting came to lower Broadway in 1825, creating a luminous spine from the Battery to Grand Street and extending east and west to the rivers the following year. Niblo's Garden, a theater at Prince Street, set gas jets inside red, white, and blue glass cups to spell out "NIBLO" when the doors

5.2

The "new Fall signs along Broadway" and Times Square featured elaborate designs and higher intensities "that pale the supreme spectacular efforts of even a season ago." "Brilliant Broadway," *Electrical Merchandising* 24, no. 3 (September 1920): 108.

opened in 1828, probably Broadway's first use of colored advertising lights.[10] Other theaters followed suit. An English visitor in 1867 took note of Broadway's lights, considering them a standard part of the "outside paraphernalia of a place of amusement."[11]

Electric street lighting was added to the mix in 1880, when Brush electric arc lamps were installed along Broadway from 14th to 26th Streets. The result was a hybrid system of electric and gas from public and private sources that brightened the boulevard.[12] As a travel guide put it, "Probably no aspect of this great city is more interesting to the stranger than that which presents itself after the gas is lighted. … From Union Square upward, Broadway is fairly ablaze with electricity and gas, massed in parterres of light at the squares and stretching away into a sparkling perspective."[13] A fad for roof gardens only intensified Broadway's lights as a number of theaters, restaurants, and clubs electrified their roof terraces in the late nineteenth century.[14] The largest of these lay atop the second Madison Square Garden, completed to the design of Stanford White in 1890. Towering over the terrace was Augustus St. Gaudens's *Diana*, ancient Greek goddess of the hunt and the Moon and one of the earliest public sculptures to be illuminated. As one account put it, she "looked down with interest on the scene, and doubtless mistook it for a patch of paradise which had suddenly been called into existence."[15]

In 1902 Stephen Crane noted along Broadway "countless signs illuminated with red, blue, green, and gold electric lamps" that drew crowds "as moths go to a candle."[16] Broadway bloomed with "multicolored bouquets of luminous advertising" for a French visitor the following year.[17] The year after that the novelist Rupert Hughes described the street as "one long canon of light" bursting with "rhapsodies of color" and "kaleidoscopes of fire."[18] By the early twentieth century, New York's Broadway was synonymous with bright lighting effects that were already an attraction in themselves (figure 5.2). The intersection of Broadway and Seventh Avenue, between 42nd and 47th streets, was Long Acre Square until 1904, when the *New York Times* moved into its new skyscraper headquarters and persuaded Mayor George McClellan to rename it. The subway, which passed underneath the square, opened later that same year. The generous volume created here by Broadway's diagonal through Manhattan's grid, along with greater pedestrian traffic from the subway and the spread of theaters along 42nd Street, helped root nighttime recreation to this spot. Broadway's proclivity for lighted signs would be intensified here over the next three decades as more than eighty theaters were built along the avenue and the surrounding streets. All would be electrically lighted.[19]

Oscar J. Gude dominated the development of Times Square's illuminated sign business in the first two decades of the twentieth century, after apprenticing as a bill poster in the 1880s and eventually starting his own company. Variously nicknamed

5.3
In a visual trope that would become commonplace, the O. J. Gude Company represented its signs along the Great White Way as a phantasmagoria. Advertisement, O. J. Gude Co., *Signs of the Times* 20, no. 82 (February 1913): 5.

the "Sign King of Times Square," the "Napoleon of Publicity," and "the Botticelli of Broadway," Gude was responsible for some of Times Square's earliest and best-known signs (figure 5.3). Besides producing them, he developed the basic business model: his fees covered façade and rooftop leases, design, construction, electrical current and operation, maintenance, insurance, and disassembly once a contract ended.[20] Many accounts credit Gude with coining the term the "Great White Way," which was used interchangeably to refer to sections of Broadway or to Times Square or to both.[21] It was used locally at first: as late as 1909 it was said to be a Manhattan colloquialism. But its use spread quickly.[22] The term burned into the public mind the image of a vivid entertainment nightscape pulsing with lights—theater marquees, street lamps, shop window lighting, carriage and automobile lights, and large, lighted advertisements. Important thoroughfares had earlier been illuminated more brightly in other cities, and lighting masts blanched streets in far smaller western towns starting in the 1880s, while expositions routinely exposed fairgoers to extravagant lighting effects. But the transmutation of an everyday thoroughfare into an enchanted nighttime "river of light," as it was frequently called, was an alchemy fixed in the minds of millions to Broadway.

By 1911 it was all but impossible to stand on New York's Great White Way and recall the former gloominess of cities after sunset, according to Elias Leavenworth Elliott, a founding editor of the *Illuminating Engineer*. Citing Robert Louis Stevenson's eulogy to gas lamplighters at the dawn of the electric age, Elliott suggested Times Square heralded the dawn of the modern world's new visual conditions and would epitomize New York City in the memories of foreign visitors upon returning to their dimmer homes.[23] Signs at Times Square only increased in size and number over the following decades, each new sign trying to outdo the last in what at least one author hinted was nearly a Darwinian competition, with the fittest sign being the flashiest.[24] As prices for electricity declined in the 1930s, Times Square became a beacon of optimism during the Great Depression. Artkraft Strauss, formed in the early twentieth century by Benjamin Strauss and Jacob Star, became the leading maker of signs, growing since then to become today a provider of signs globally. The company had an early franchise for neon light technology from Georges Claude, the inventor of neon tube lighting, and was a favorite of sign salesmen. Foremost among the salesmen was Douglas Leigh, who, according to a frequently told story, arrived in New York around 1930 with just a few dollars to his name. By 1933 he was advertising A&P's Eight O'Clock brand coffee with a twenty-five-foot cup at the corner of 47th Street and Seventh Avenue that unfurled ribbons of real steam. Within a short time he owned leases for many of the most prominent rooftops around Times Square and had proposed some of Broadway's most memorable signs. The journalist and radio broadcaster Walter

Winchell wrote in 1935 that the "outstanding man, the one who knows more about Broadway signs and bulbs than anyone," was Leigh, who was twenty-five at the time and nicknamed the "Boy Sign King."[25] As the architectural and urban critic Paul Goldberger wrote years later, Leigh "gave nighttime New York its visual identity for a whole period of our history."[26]

5.4
A two-story pickle in flashing green lights. H. J. Heinz electric billboard, Cumberland Hotel, 23rd Street and Broadway, New York. O. J. Gude Co. and Artkraft Strauss, ca. 1898.

MOVING WORDS

Wherever possible, motion has been the magnet used to attract the attention of the public.
Frank C. Reilly, 1912[27]

In numerous accounts, the animation of Times Square's giant signs distinguished the square from the rest of Broadway.[28] The illusion of motion, however, had long been a feature of signs. The flickering wicks of oil lamps threw dancing shadows across ancient shops at night, and magic lanterns projected moving images to attract customers as early as the sixteenth century. A sheet of paper oiled to translucency could be set on a rotating wheel and backlit to make an especially eye-catching display. Even the slight trembling of gas jets in the early nineteenth century was part of the fascination of lighted signs.[29] The first electrically illuminated sign was a moving one. Created in 1882 by Thomas Edison's top engineer, William Joseph Hammer, to display at the Crystal Palace in London, it was hand-operated to spell out "Edison," one letter at a time, with incandescent bulbs. A flashing sign with four-foot-tall letters at the 1893 World's Columbian Exposition in Chicago was a popular attraction, even though its electrical contacts threw off sparks and an attendant was hired to stand guard in case of fire.[30]

New York City's first electric sign also flashed letters on and off, blinking at Broadway and 23rd Street on the blank north wall of the Cumberland Hotel. The industrialist Austin Corbin had leased the tall surface for painted signs. In 1892 he installed an electric sign there to advertise houses for sale in Manhattan Beach, Brooklyn, which had the added—perhaps the primary—benefit of increasing traffic on the Long Island Rail Road, which he operated as well.[31] The novelist Theodore Dreiser, when he first saw that sign, admired its choreographed illumination: "As one line was illuminated the others were made dark, until all had been flashed separately, when they would again be flashed simultaneously and held thus for a time." He knew of similar displays in Atlantic City and Coney Island, "but this blazing sign lifted Manhattan Beach into rivalry with fairyland."[32] All of Madison Square became "a brilliant example of all the latest 'wrinkles'" in electric lighting, including the lights of Madison Square Garden, Corbin's sign, and another large sign whose "letters of fire ... proclaim the fact that a new brand of cigarettes has captured the town." After Corbin's death four years later, the *New York Times* seized the opportunity to lease the wall for its own advertising. The Heinz Company leased it afterward and hired Gude to fashion an enormous pickle outlined with green-tinted bulbs that flashed on and off (figure 5.4).[33]

Flashing technology advanced rapidly in the early twentieth century. Around 1900 Egbert Reynolds Dull founded a company to capitalize on his serendipitous

5.5
With motion a key desideratum, sign makers developed numerous ways to create animated letters. C. A. Atherton, *Electrical Advertising* (Cincinnati, OH: Signs of the Times Publishing Co., 1925), 85.

adaptation of bimetallic thermostats for use in lamps to create the first automatic flasher, obviating attendants or customized motor setups. Dull is also credited with having come up with the term "flasher," for the new device, drawing on the earlier use of "flash" to denote a brief burst of light (figure 5.5).[34] The "motograph" was a particularly important breakthrough in enlivening text. It was based on a patent filed in 1911 by Everett Bickley, a twenty-three-year-old who would go on to become a prolific inventor.[35] Prior to this time, the "talking sign," as it was known, consisted of a series of bulbs set in metal troughs that visually joined to create letter forms when the bulbs were switched on. Rows of troughs could be set up to spell out words.[36] Changing the text in these signs was laborious, however, and the motograph was a considerable improvement.[37] It consisted of an array of potentially thousands of lamps, each wired to an electrical contact that was in turn regulated by an insulating perforated ribbon continuously rolling over a wired metal sheet, analogous to a player piano's operation. The manufacturer provided clients with custom ribbons since the perforations were just holes punched in paper and could reproduce virtually any type style or symbol.[38] Even better, changing the text was as easy as changing the ribbon, which could be done in a minute and without any rewiring.[39] Finally, the system was inexpensive to operate since only about one quarter of the lamps were lit at any one time.[40]

The Bickley Motograph was "without question the livest, most attractive and modern flashing sign of the age and the biggest step forward into the future ever taken," hyperbole being no defect in the outdoor advertising industry.[41] Its effects became the source of considerable fascination, described in great detail in trade periodicals, with special attention given to the convincing illusion of moving letters. It appeared "as if the entire word had been carried across in solid illuminated letters, but in reality the lamps do not move," explained one such account.[42] More important, the effect fulfilled the prime objective of outdoor advertising, it guaranteed people would notice: "The first thing that impresses the observer is the fact that the advertising matter *is in motion*. Never has a display been developed that embodies this principle, or so abruptly arrests and completely absorbs the attention of all beholders."[43] Standing out from dozens of other sign controllers, the motograph spurred further innovations centered on making text move.

The motograph also inspired visions of a way in which architecture and moving words might merge. In 1912, Frank C. Reilly, a street car electrician who later designed signs around Times Square, imagined bands of text traveling around the skyscrapers that were just then sprouting in Manhattan. Stirred, he filed a patent the next year to refine the motograph.[44] Fifteen more years would pass before he helped create the best-known motograph, the *New York Times*'s news "zipper," which began operation in November, 1928 (figure 5.6). Wildly popular,

5.6

The New York Times "zipper." A new kind of urban reading was pinned to a new urban building type. Farm Security Administration/Office of War Information Collection, Library of Congress, Prints and Photographs Division.

the zipper wrapped the *Times* tower with five-foot-tall letters composed of nearly 15,000 bulbs and running about 360 feet. Western Union transmitted news bulletins by telegraph to a control room inside the tower, where it was transcribed into points of light, "the news of the world flashed out in electric letters, readable a mile away."[45] A good deal of the zipper's hold on the imagination owed to its wrapping a modern means of communication around a modern form of construction, in this manner merging a technology that, by dispensing news to thousands of readers regardless of where in the square they stood, seemingly collapsed time and geography with a technology that, in its speed and scale of construction, seemed to defy gravity. Together, these two technologies positioned New York at the center of a new kind of urban modernity in which live updates from afar shattered the site specificity generated by the city's uniquely tall towers.

MOVING PICTURES

Fact is, the whole thing is only one step removed from
the movies anyway.
***Tide*, 1937**[46]

Bottles of beer appear on the firmament and transform
themselves into dwarfs drinking; showers of gold peanuts
fall from the skies; dragons breathing smoke become
a film title; cigarettes are ignited; automobiles materialize.
Mountains, towns, lamaseries, men with top hats,
nude women with teeth, spring into existence on the
facades and are wiped off into oblivion.
Odette Keun, 1939[47]

As exciting as it was to outdoor advertisers, moving text took second place to moving images in capturing viewers' attention. Moving pictures broadened what might be called the bandwidth of a sign: just as a picture is worth a thousand words, ran one comment, "one animated display does the work of 1000 pictures."[48] Flashing images, in other words, were the information-dense pole on the advertisers' spectrum of meaningful communication. Thousands gathered nightly to stare for minutes on end at the simple animations of the fifty-foot-tall "Miss Heatherbloom" (figure 5.7). Built around 1905 to advertise petticoats, she stood unembarrassed before her onlookers while an invisible wind lifted the hem of her skirt. The sign reached a new "realism in electric display" and, as the first flashing female flasher, set an erotic standard for Times Square that would become more explicit in years to come. The 1912 Corticelli Kitten was a naughty tabby vainly trying to tangle a spool of silk, with "the spool and the kitten's paws being in continuous motion."[49] A White Rock Table Water sign from around 1915 depicted streams of

5.7
The wind-lifted petticoat of Miss Heatherbloom captured the attention of thousands of visitors nightly in Times Square. Image AAA0109, reproduced with permission from the OAAA Digital Archives Collection, Rubenstein Rare Book and Manuscript Library, Duke University, Durham, North Carolina.

colored water pouring from lions' mouths bracketing a clock with moving hands that New Yorkers checked against their own wristwatches. The 1917 Wrigley's Spearmint Chewing Gum sign was festooned with figures of feathery peacocks, fountains and foam, and two formations of turning "spearmen," all of it fitted into a filigreed frame (figures 5.8 and 5.9). Wrigley's extolled the sign as much as the product, boasting of hundreds of thousands of people coming to Times Square to see "the largest electric sign in the world," replete with dancing figures.[50] All were made by the O. J. Gude Company, which was quick to use the new flashing technologies as they became available and seemed to understand that lively, lighted advertisements were not just eye-catching. They contributed a new sort of visual modernity to the city.[51]

The best-known sign to use pictorial effects to draw attention to an advertising message was the "Leaders of the World" sign, built by the Rice Electrical Company in 1910 on the roof of the Hotel Normandie at Broadway and 38th Street, facing south toward Herald Square (figure 5.10). It featured an animated Roman chariot race that seemed like "real life-action in fire."[52] Above the race a band of text presented an advertiser's message. In industry parlance, it was a "flash light" sign: an outdoor advertising display running serial promotions presented at a frequency and duration established by contract. The race, repeated on a thirty-second loop, held the crowd's attention while the advertisement flashed just above. It was an embryonic instance of a new kind of urban reading that exploited the eye's natural responsiveness to motion in order to smuggle an advertisement into the spectator's view. To borrow Max Horkheimer and Theodor Adorno's analysis of the commercial motivations of the culture industry, the "Chariot Race," as it was often called, traded a brief entertainment for viewers' attention, which, lost in a moment of leisure, was yoked to commercial promotions.[53] In this sense, the Chariot Race was a blank page prepopulated with eager readers unwittingly readying themselves to receive whatever might be spelled out before them.

By the 1920s and 1930s, the sign makers' palette was virtually unlimited, with "all types of devices for compelling attention," creating illusions, and generating moving words and images.[54] By all accounts, Douglas Leigh was the master, bringing motion into the third dimension with a ring-blowing Camel smoker, a bubbling detergent, a block-wide waterfall, and a building-sized snowflake. His unbuilt designs for a "helium-filled plastic orange dripping simulated juice" to advertise Tropicana products, a five-mile-high swinging searchlight for Eveready batteries, and a small fleet of lighted blimps for any advertiser, to name only a few, are unstudied landmarks of an incipient media architecture (figure 5.11). In 1937 he introduced a new level of animation with the Epok system, which made use of photocells connected to a grid of light bulbs. Although it was an Austrian invention with the

5.8

Boasting more than 17,000 lamps, Wrigley's pitched its chewing gum on the basis of sponsoring the world's largest electric sign. Illuminated electric sign for Wrigley's gum, Times Square, New York City, ca. May 1917. William D. Hassler Photograph Collection, image 34353. New York Historical Society. Photography@New-York Historical Society.

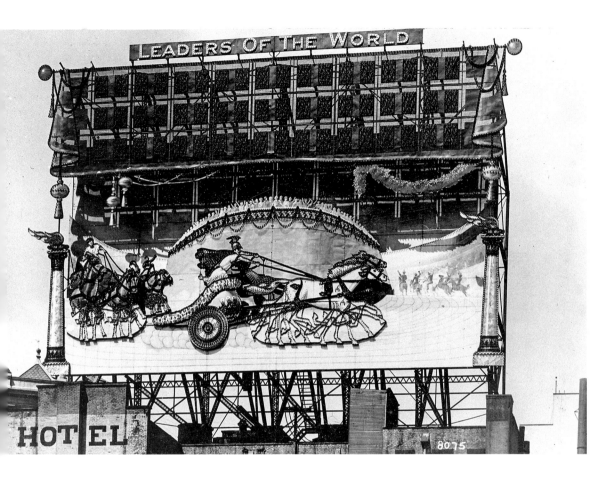

5.9

For years, big, bright, novel, and visually compelling signs were reported as news and described by their dimensions, the number of bulbs operated, the miles of wires used, and the number and sophistication of the flashers, comparable to the competition to build the world's tallest skyscraper. The 1910 Knights Templar sign in Chicago was said to be the tallest spectacular in the world at the time. Knights Templar sign, postcard. Hilton Litho Co., Chicago, ca. 1910. Collection of the author.

5.10

After it was completed, the Chariot Race became a touchstone for "flash light" signs. Captioned "Leaders of the World," Rice Electric Displays, 1910. Image AAA9904, reproduced with permission from the OAAA Digital Archives Collection, Rubenstein Rare Book and Manuscript Library, Duke University, Durham, North Carolina, ca. 1940s.

ELECTRIC SPEECH IN THE CITY

5.11

Of the many schemes over the years to remake Times Square, Douglas Leigh's remain the most radical. *BoxOffice*, Jan. 25, 1947, cover. Douglas Leigh Papers, 1903–1999, bulk, 1924–1999. Courtesy of the Archives of American Art, Smithsonian Institution.

parent company headquartered in Stockholm, Leigh's use of the technology was unique. "Europeans don't seem to be ready for spectacular advertising yet," he guessed.[55] Leigh refined the device with his engineer, Fred Kerwer, and hired Otto Messmer, the animator best known for creating Felix the Cat, to design image sequences.[56] With this system, actors' shadows could be cast against the photocells for a live-action performance. Leigh's first Epok sign was greeted enthusiastically: "Everywhere New York City people are heard talking about the animation feature"[57] His second, for the musical *High, Wide, and Handsome*, then playing at the Astor Theater in Times Square, was a miniature of the film and earned its own review in the *New Yorker*.[58] As Leigh noted, unlike other spectaculars, which were limited to short display cycles, the Epok sign could go on as long as the film it was based on.[59]

5.12
A kind of cinema of the streets was created with magic lantern projections on surfaces. Evening of Nov. 6, 1888, Madison Square, New York. *Harper's Weekly*, Nov. 17, 1888, 40.

Such signs created a cinema of the streets. *Son et lumière* spectacles had brightened façades in the eighteenth century and magic lanterns projected advertisements onto buildings in the nineteenth century, although these were limited by the brightness of available light sources. The stereopticon, the outcome of several technical advances in the nineteenth century, made outdoor projections more viable. They notably served the advertiser's need both to attract attention and to deliver a message. In 1871, a Mr. Van Dusen stopped passersby "in the open air, by means of mammoth views, interspersed with business cards. The views are so pretty that the public is willing to stand the advertisement in order to see the whole of the views. The idea is novel."[60] Projected election results were perhaps the greatest means of creating a crowd, and newspaper publishers staged a number of such events (figure 5.12). These were episodic, however, and, though often revelrous, nominally civic.[61]

Anticipated by these earlier outdoor projections, Times Square was different in that it was fixed in location and dominated by commercial rather than political speech. Times Square even resembled an arena, though inverted, with the audience at the center and the performance along the periphery. Comprised of two long and open scalene triangles, their acute angles touching along the oblique line of Broadway, and walled by theaters, retail establishments, and office buildings, it was especially well suited for squinting at signs. From its center at the intersection of Broadway, Seventh Avenue, and 45th Street, and looking either north or south, the space opened up like the field of vision itself. With signs situated on façades or rooftop scaffolds, views were encircling and unimpeded, a multiscreened, moving palimpsest of overlapping planes of flashing lights. Times Square was likened to "a vaster better-lit stage" than could be found in any of the theaters lining it. Indeed, in many accounts, standing and staring skyward was the more diverting part of an evening out in New York.[62]

Moreover, unlike other New York City landmarks, Times Square changed from moment to moment, as signs flashed at varying intervals, and from season to season, as leases expired and new signs were put up. Whatever the consistency of any one sign, the array was plotted by an impenetrable calculus of available rooftops and façades, leasehold terms, flasher techniques, and sign designs. Different stories, so to speak, were recited as viewers took in details of the scene in diverse ways. Summarizing the typical tourist's response, the mystery writer and *New York Times* reporter Stephen Chalmers remarked, "This is certain—no one ever strolled up Broadway under the white lights but longed to do it all over again," despite the headache Chalmers assured readers would follow after an evening staring at flashing signs.[63]

Perhaps the most frequent cliché regarding Times Square's visual stories was that they were more captivating than theater or cinema and, even better, free (figure 5.13). Gude had already called Times Square "a great free exhibition for strangers from all over this country and Europe," declaring that tourist "coaches nightly take strangers up Broadway to see this phantasmagoria of lights and electric signs."[64] The Chariot Race sign—more realistic "than the finest coloured cinematograph picture"—was seen, it was said, by more people "than any other 'free show' of the kind in the world." Squads of policemen were summoned to keep the crowds contained as they gawked, transfixed by the animated scene.[65] All of Times Square was dubbed "the finest free show in the world."[66] The trope of a free show only amplified over the years as effects became more naturalistic. "Crowds packed the sidewalks facing the display until it is impossible to pass" in order to see Leigh's first Epok sign, for Trimble whiskey.[67] In 1945, to promote a war bond drive, Leigh staged a live-action Epok reenactment of the flag-raising at Iwo Jima. It was a "a brand-new show on the Great White Way, and like so many marvels of Broadway, it's free. … Of course, there are no reserved seats, and you have to crane your neck, with the crowds, but it's really worth seeing, even with standing room only!"[68]

5.13

The electric sign as an evening's entertainment. Captioned "Will Run All Season," *New York World-Telegram*, n.d. Signed Johnstone and Jim Crouch. Douglas Leigh Papers, series 10, reel 5844, p. 63. Courtesy of the Archives of American Art, Smithsonian Institution.

The lifelike aspect of such performances, always part of their attraction, added a twist to Horkheimer and Adorno's 1944 claim that contemporaneous naturalistic tendencies in film blurred the line between corporate production and lived experience. Leaving the cinema, they wrote, the moviegoer "perceives the street outside as a continuation of the film he has just left," with the perceptual continuity consequently confusing the distinction between the private space of the cinema and the public space of the city street. With the free shows put on at Times Square, the "street outside" duplicated the visual experience of a private cinema, but it did so within the public realm of the city square. The signs reproduced on the street the cinema that, according to Horkheimer and Adorno, reproduced the world outside.[69] Illuminated and animated commercial speech became in this way a means of colonizing the public sphere by private interests and transforming an urban space into an arena for visual pleasure.

MOVING GOODS

It has come to be recognized that light in itself is one of the greatest of advertising forces in modern business.
Dry Goods Reporter, 1915[70]

Visual movement, whether word or image, was widely believed by industry spokesmen to make advertisements hypnotizing and their deployment in urban space more effective. Admen translated the eye's natural sensitivity to motion into promotional merit, straining at times, absent actual evidence, to enumerate it: "It is claimed that a flashed electric sign easily possesses 20 times more advertising value than one burning steadily," ran a 1910 assertion, guaranteeing it would be "a handsome dividend payer."[71] Flashing signs recapitulated a fundamental truth about most outdoor advertising: people saw it briefly, usually in passing. Flashers shortened viewing time even further, to the split second. By compelling attention, flashing signs appropriated the visual volition of passersby—the control of their own gaze—and forced notice of the message. Their special appeal, advocates suggested, was that the human brain was wired to respond to brightness and motion.[72] They effectively "monopolize the field of a buyer's consciousness" and elbow out competing appeals to the consumer's attention.[73] Energetic effects such as wriggling snakes and running borders ensnared passersby and transfixed the tourists who sought them out deliberately.[74]

The psychology of attention had been studied for decades by this time, but the field's aura of empiricism had only recently attracted the advertising industry. Writers on the subject began talking regularly about the "attention value" of a sign. Regarding the Bickley Motograph, for instance, Reilly said that "the effect produced, the 'attention value,' is so strong" that two serious road accidents were only narrowly avoided in Detroit, where it was first installed.[75] The problem, as it came to be understood, was that moving pictures attracted attention effectively but the advertising message was borne by text. Thus, for effective design of a sign, the problem was to "transfer ... attention from the automatic part of the display to the advertising matter." It was a vicious cycle, though, since "unconsciously the mind and the eye wander back to the moving part of the display" and thus failed to fully absorb the ad's main message.[76] To be fully successful, the "attention value" of a sign had to be paired with its "memory value," that is, the likelihood a viewer would retain the promotional message.

Together, the mental faculties of attention and memory underpinned what might be called a theory of motion in advertising. These factors had been discussed in early twentieth-century psychology journals, but the most influential work in this regard was Hugo Münsterberg's *Psychology and Industrial Efficiency*, from 1913. Münsterberg was a German-American pioneer of applied psychology who studied at Leipzig in the 1880s with the pioneer physiological psychologist Wilhelm Wundt. He was squeamish about taking promotional come-ons seriously but justified his work by noting that advertising had become a substantial expenditure for many businesses and thus warranted closer examination. His study centered on the various "mental effects" of advertisements, including attention value, something like a knock on the door of a consumer's awareness, and memory value, which evoked "a warm feeling of acquaintance" that could influence purchasing behavior.

Münsterberg described the visual attributes that stimulated attention and memory, such as legibility, vivid impressions, originality or unusual features, strong colors, interesting design, and robust associations. These factors combined in various ways to trigger a range of basic emotions, including curiosity, sympathy, and even antipathy. Equally important was repetition, which affected attention and memory alike by inaugurating a cycle of alertness as an advertisement etched itself in memory. Repetition, he wrote, "awakes the consciousness of recognition, thus exciting the attention, and through it we now turn actively to the repeated impression which forces itself on memory with increased vividness on account of this active personal reaction." Granting that the psychology of advertising was still in its infancy, Münsterberg pointed out that admen were oblivious to basic psychological principles.[77] This was plain to see, he said, in the "antipsychological absurdities which any stroll through the streets of a modern city forces on us."[78] Münsterberg does not mention wigwagging electric signs, although one might reasonably presume that was exactly what he had in mind.

Later researchers took up Münsterberg's challenge to develop a psychology of advertising, with attention value and memory value as core concepts (figure 5.14). As the number, size, sophistication, and omnipresence of electric signs continued to grow, these became the leading terms employed to explain their effectiveness. Henry Foster Adams, an empirically minded psychologist who came to specialize in advertising, described the visual factors that "flood the cortex with nervous energy, thereby compelling attention."[79] Electric signs excelled at generating such nervous stimulation, he said, but they often did so at the expense of memory value. He recalled in particular "an electric sign on Broadway … of the variety which gives the illusion of movement." It grabbed his attention, but the visual activity was so intense he became dizzy and was unable to read the text; it had "tremendous attention value" but left no imprint on memory.[80] Others were more open to spectaculars and sometimes singled out Times Square as the Colosseum of epic displays of attention value and memory value. Harry Tipper, an advertising executive

5.14

Elements that made large illuminated signs attention-getting and memorable included individuality, size, brightness, and repetition. W. C. Brown, *Electrical Advertising: Its Forms, Characteristics, and Design* (Cleveland, OH: National Lamp Works, General Electric Co., 1927), n.p. Reproduced with permission from the Archives and Research Center, Museum of Innovation and Science, Schenectady, New York.

with the Texas Company, forerunner to Texaco, suggested that electric signs were a form of "complete advertisement," his formulation for a promotion that attracted and held attention, created associations that precipitated memory value, and ultimately inclined potential customers toward preferring one brand over another.[81]

The conceptual pair of attention and memory became fixed as a metric for assessing illuminated signs, with many in the field agreeing that animated electric signs were, by the industry's own criteria, the most effective and swiftest means of attracting notice, conveying meaning, and taking up residence in memory. As Leigh summarized, "a good sign has two aspects, and only two. It must attract attention, and it must have memory value."[82] Scale and brightness drew viewers, he said, "like moths to a flame." But moving things settled into memory. Recounting his own inventions, he asserted: "The more animation in the display, the more memory value it contains."[83] He went even further to suggest that such effects could displace older memories to become the primary figure when, for instance, an individual imagined rising smoke.[84] In this formulation, Leigh suggested a model of commercially conditioned ideation wherein memories gained from firsthand experiences could be altered by repeated encounters with promotional forms.

Many agreed. The American mind-set was characterized by "restless, nervous energy," in the historian Frederick Jackson Turner's formulation, and hungry for new experiences and expanded fields of action. Dynamic signs both reflected and further reinforced this "modern desire for action," according to a representative claim in an outdoor advertising trade periodical.[85] "Our high-tension living is being visibly expressed in modern sign and outdoor advertising displays. Their vivacity is in key with tastes of a public that is constantly keyed to a high pitch in constantly going places and doing things."[86] In these and similar observations, spectacular signs were as much expressions of a national personality centered on novel experiences as they were promotions for manufactured goods.

READING LIGHT

It makes your head reel, for that blaze and riot of light isn't static;
it flares, flows, writhes, rolls, blinks, winks, flickers,
changes color, vanishes and sparkles again before you can
open your mouth to gape.
Odette Keun, 1939[87]

In becoming a singular space of illuminated commercial speech, Times Square also became a unique place for urban reading (figures 5.15 and 5.16). Signage, previously a supplemental and for many a corrosive element of the streetscape,

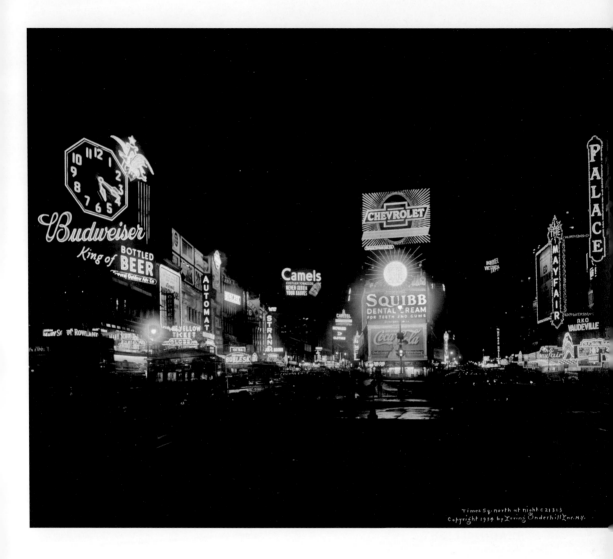

5.15

An arena for reading in public. *Times Square North at Night, New York City,* photograph, ca. Jan. 16, 1934. Library of Congress Online Catalog, Prints and Photographs Division (https://www.loc.gov/item/2006677804).

was foundational there and, in its way, constructive. Advertising, still a suspect cultural practice—hardly a generation away from snake oil salesmen—became thunderously thrilling at Times Square, a former background newly flowered into a foreground-filling figure. Whatever a sign's message might be, the flashing words appeared more akin to a graphic performance than to textual reference. Times Square became so much a place for "gyrating sky signs" that in many travel accounts the signage eclipsed the site's other attractions.[88] People came, looked up, and spun around to take in the bowl-like space studded with blinking messages tumbling forth from all sides simultaneously, each jostling the others in a frenzied effort to reach and command visitors' attention and inhabit their memory.

5.16

The lesson of Times Square was to imagine the urban sphere of reading and to create a sign with "a range of effective reading distances." "Circulation Measured in Dollars Essential to Sign Design," *Signs of the Times* 45, no. 1 (September 1923): 30.

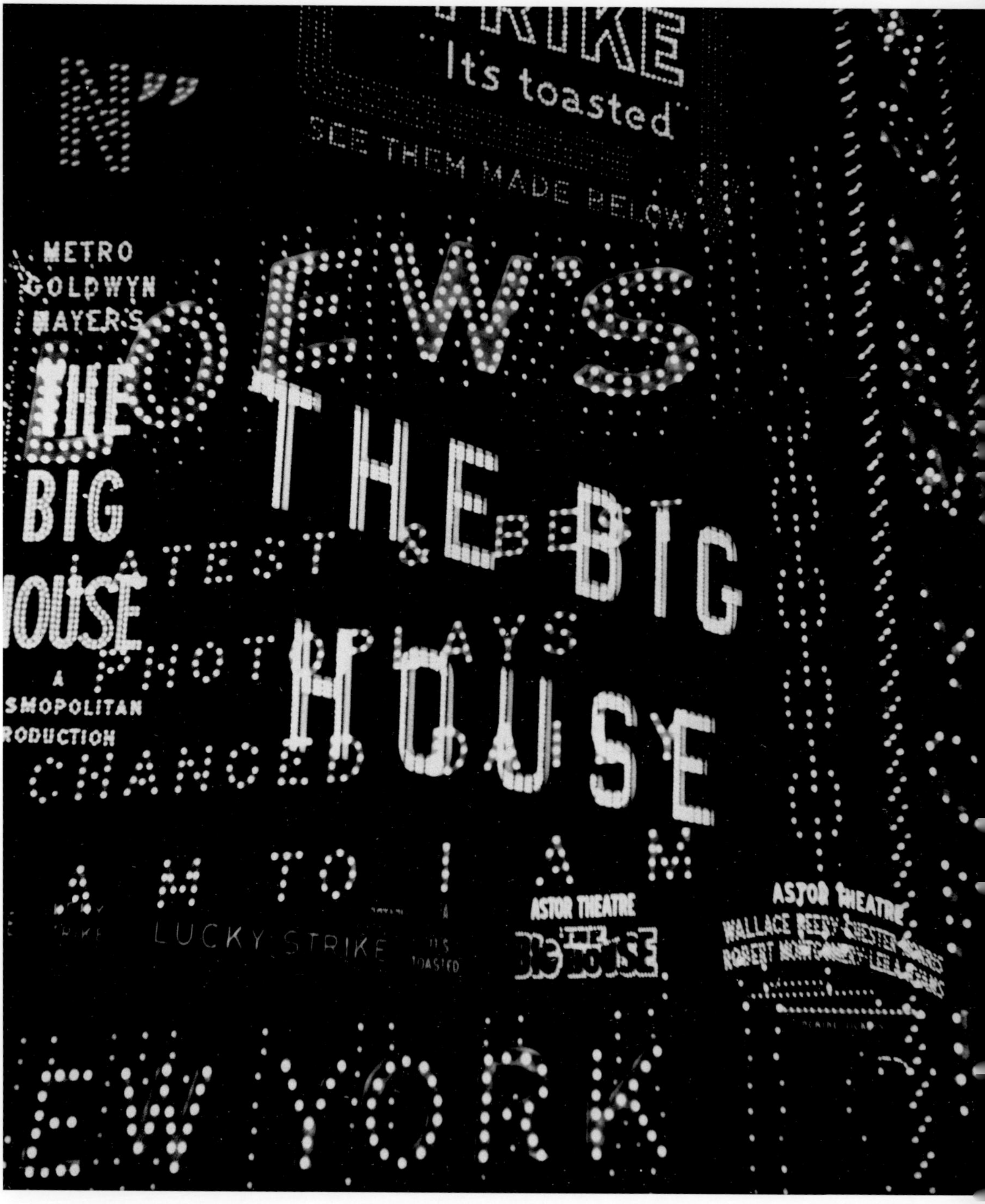
Its toasted
SEE THEM MADE BELOW
METRO
GOLDWYN
MAYER'S
THE BIG HOUSE
A
PRODUCTION
LUCKY STRIKE
ASTOR THEATRE
ASTOR THEATRE
WALLACE
YORK

BROADWAY BABBLE

The advertising value of a sign is proportional to the number of people who read it.
Signs of the Times, 1920[89]

The compulsory aspect to electric signs that so many people experienced in Times Square was understood by outdoor advertisers to be the medium's most powerful and validating attribute. No one expressed the idea more succinctly than Gude himself, who wrote: "The fundamental principle of outdoor advertising is that 'he who runs MUST read.'" The phrase, possibly an inversion of an earlier lament that few had time to read thoughtfully in the harried modern world—he who reads must run—was common to admen. While other sorts of advertisements required a viewer's cooperation, "the outdoor advertising sign asks for no voluntary acquiescence from any reader. It simply grasps the vantage point of position and literally forces its announcement on the vision of the uninterested as well as the interested passerby."[90] In this and many other accounts, sophisticated displays of light, color, text, and visual diversity entangled the mind in its effort to decode linguistic symbols. Put another way, the more that signs shouted, the less viewers read, at least voluntarily. Times Square's "electric scream," as one observer termed it, could not go unheard.[91]

This crazily punctuated model of reading helps explain one of the most consistent and striking responses to Times Square: feeling overwhelmed by the signs and, in trying to describe the experience, finding oneself merely able to list them one by one, as if to confirm each sign's individual logic and at the same time demonstrate their topical disjointedness as a collective (figure 5.17). A German visitor, for example, writing about theater in 1922, stopped to look at and list the lighted advertisements: "Here flies a bird of electric lights, there champagne spurts from a bottle, a Scotsman dances a jig to advertise whisky, here again a jet of waterfall foams, a red ball shoots from a cannon, a green squirrel speeds round a circle."[92] The list of such lists of Times Square's signs is a long one.

5.17
Visual artists, including Jaromir Funke, Man Ray, Fritz Lang, and El Lissitzky, tried to capture the dynamic and disorienting aspects of flashing signs on Broadway and Times Square with hallucinatory images. Likewise, industry advocates assembled hallucinatory collages of signs to illustrate their articles.
Here, Walker Evans's view of Times Square, ca. 1930. The J. Paul Getty Museum, Los Angeles. Walker Evans, Times Square/Broadway Composition, 1930, gelatin silver print, 27 × 23.5 cm. © Walker Evans Archive, The Metropolitan Museum of Art.

The English writer Arnold Bennett was especially awed by Times Square and acutely sensitive to the linguistic paralysis he felt looking up at signs. Above the street-level lights, he wrote in 1912, was a sliver of night,

> "And above the layer of darkness enormous moving images of things in electricity— a mastodon kitten playing with a ball of thread, an umbrella in a shower of rain, siphons of soda-water being emptied and filled, gigantic horses galloping at full speed, and an incredible heraldry of chewing-gum. ... Sky-signs! In Europe I had always inveighed manfully against sky-signs. But now I bowed the head, vanquished. These sky-signs annihilated argument. ... I was overpowered by Broadway. "You must not expect me to talk," I said.[93]

Dumbfounded, he regressed to a child's grasp of the world, left only to name what he saw and, following that, saying no more. All the virtual motion disrupted his emotions and disabled higher functions. A novelist, no less, he was left speechless by the place.

In 1914 the American writer Rupert Hughes described his protagonist, Lieutenant Forbes, who has just returned to New York after some absence:

> Forbes sat in the dark room in an arm-chair and muffled his bathrobe about him, watching the electric signs working like solemn acrobats—the girl that skipped the rope, the baby that laughed and cried, the woman that danced on the wire, the skidless tire in the rain, the great sibyl face that winked and advised chewing-gum as a panacea, the kitten that tangled itself in thread, the siphons that filled the glasses — all the automatic electric voices shouting words of light.[94]

Forbes had nothing to say in response to the "words of light" shouted at him by "electric voices" since, it now seems clear, silence was the other half of one's dialogue with Times Square. Indeed, he had taken care to muffle himself as he sat to see what the signs had to say.

In 1916, Rupert Brooke, another English writer, likened the sky-signs to a sky-high pantheon of enigmatic gods, whose activities thundered over the Earth below but whose import forever escaped us mortals. He listed them: a devil unable to bend back the bristles of "vast fiery tooth-brushes"; opposite, "a divine hand writing slowly ... its igneous message of warning to the nations: 'Wear—Underwear for Youths and Men-Boys'"; nearby, "a celestial bottle, stretching from the horizon to the zenith." Not far from these "a Spanish goddess, some minor deity in the Dionysian theogony, dances continually, rapt and mysterious. ... And near her, Orion, archer no longer, releases himself from his strained posture to drive a sidereal golf-ball out of sight through the meadows of Paradise; then poses, addresses, and drives again." For all their determined activities, these "coruscating divinities," including two warring youths, "clad in celestial underwear," remained a mystery: "What gods they are who fight endlessly and indecisively over New York is not for our knowledge." One figure in particular, a colossal woman's head, seemed to Brooke poised to lend meaning to the scene, to offer "the answer to the riddle." Brooke looked past the sign's recommendation of chewing gum to pursue a deeper meaning in the woman's winking eye, which ultimately proved to be an "answer that is no answer" to his quest. This "Queen of the night," as Brooke called her, was a sphinx, a reticent custodian of meaning: "And the only answer to our cries, the only comment upon our cities, is that divine stare, the wink, once, twice, thrice. And then darkness." She ruled a *flammantia moenia mundi*—a fiery-walled world—and in response to a search for sense added word to image only to display the incongruence of their sum.[95] Although his words were satirical, Brooke had pinpointed the paradox at the heart of Times Square's spectacular signs: the marshalling of considerable technological resources to monumentalize banal things.

As another English writer, G. K. Chesterton, wrote in 1922, the very desire for meaning to emerge from these overscaled signifiers only diminished their overpowering effect, and ought therefore to be renounced. Comprised of "the two most vivid and most mystical of the gifts of God; colour and fire," the signs "hung in the sky like a constellation," at first appearing to proclaim a paramount principle of the entire nation. But they did not. Of Broadway's signs, he said: "What a glorious garden of wonders this would be, to any one who was lucky enough to be unable to read." Only an illiterate could appreciate "sights almost as fine as the flaming sword or the purple and peacock plumage of the seraphim."[96] In such accounts, visitors were advised to surrender any impulse to find narrative, order, or proportion and accept instead the primitive linguistic state that Times Square ushered them toward. It was a place where one went to see in staccato and was left able only to say in sequence what one saw. Beneath it all, however, the compulsion to read remained, despite the impossibility of ever assembling a narrative. In this, Times Square heralded the emergence of a new urban semiotic regime whose colossal signifiers contended with each other to outshout their trifling signifieds to stammering onlookers below (figure 5.18).

THE FLASH AND THE LIST

The value of such an advertising medium ... [is] mainly dependent upon
the use of the aforesaid intervals of darkness between
the displays so as not only to compel attention but to avoid the
confusion and blurring effect that would result if the displays
melted into each other without the pauses caused by such intervals
of darkness, that is to say, without the distinct use of the "flash."
Park & Tilford v. Realty Advertising, 1919[97]

5.18

A photograph less of Times Square itself than of the experience of a new type of urban reading. The accompanying caption describes the frenzied psychic effect of the flaming script and rocket-fire of moving illuminated advertisements as their light erupts and vanishes and erupts again over thousands of cars and eddies of humanity. The result is a deep and awesome beauty still in formation. Fritz Lang, photographer. Erich Mendelsohn, *Amerika: Bilderbuch eines Architekten* (Berlin: Rudolf Mosse Buchverlag, 1926), 150.

As a literary device, the list predates the flasher. Nonetheless, it rhetorically repeated the action of the flashing sign: illuminated content, whether meme, letter, or entire message, followed by an interval of darkness to set up the succeeding flash of light. The unlit interim between flashes works like the comma or line break in a list. Each entry is self-contained, without any necessary connection to what comes before or after. On the one hand, inclusion in the list is what initiates the need to compare what might otherwise be incompatible things. Their separation, on the other hand, is what continuously defers comparability and makes possible the linking of unlike elements.[98] Michel Foucault addressed the limits of such conventional systems of classification in a reference to a short 1942 essay by Jorge Luis Borges. In "The Analytical Language of John Wilkins," Borges describes a fictitious Chinese encyclopedia, titled *Celestial Empire of Benevolent Knowledge*, that contained an absurd taxonomy: "animals are divided into: (a) belonging to the Emperor, (b) embalmed, (c) tame, (d) sucking pigs, (e) sirens, (f) fabulous, (g) stray dogs, (h) included in the present classification, (i) frenzied, (j) innumerable, (k) drawn with a very fine camelhair brush, (1) et cetera, (m) having just broken the water pitcher, (n) that from a long way off look like flies."[99] For Foucault, the power of Borges's catalogue of encyclopedia entries resided not in the fantastic qualities of the categories themselves but in "the narrowness of the distance separating them. ... What transgresses the boundaries of all imagination, of all possible thought, is simply that alphabetical series (a, b, c, d) which links each of those categories to all others." Foucault added, "The sudden vicinity of things that have no relation to each other; the mere act of enumeration that heaps them all together had a power of enchantment all its own." But the genuinely "monstrous quality" of Borges's list was the sheer incommensurability of entries that were nonetheless united in their simultaneous presence in the form of a list.[100]

Advertisers understood perfectly the pivotal role of gaps between presentations of unlike content. Each flash of a sign must be completely disjoined from the prior flash. If not, then messages might be muddled and the matter might become the basis for a lawsuit. In a case that looped through New York State courts from 1913 to 1915, a candy retailer sued a sign maker for failing to provide the contracted number of flashes stipulated and, more telling, for misrepresenting the effectiveness of the sign, put up at Broadway and 47th Street. Despite their differences, plaintiff and defendant agreed that a flashed message comprised two distinct elements: the actual burst of light and a preceding moment of darkness. They discussed in detail just how long the "appreciable interval of darkness" ought to be. Indeed, the language approached clinical precision, with "appreciable," "interval," and "darkness" separately analyzed in turn. As the plaintiff's attorney put it,

> This "flash" is of very great value in advertising, especially in a locality where many advertisements are displayed continuously, producing a great volume of light.

A steady display, uninterrupted by intervals of darkness, only adds to the general illumination. On the other hand, the sudden appearance of a new display after an appreciable interval of darkness produces an impression abruptly gained, and more quickly attracts the attention of the passerby.[101]

In other words, the parties to the suit knew well that the eye needed time to adapt to the relative darkness if the subsequent flash of light were to stand out and draw attention to itself. As the plaintiff's attorney stressed, "The psychology of the case also shows that if you don't have an appreciable interval of darkness the efficacy of the sign is destroyed."[102] The plaintiff argued that it had purchased both the flash of light and the interval of darkness and inadequate amounts of both had actually been provided. Plaintiff's and defendant's shared standard of excellence in "flash light" signs was the Chariot Race, just a few blocks away. As all parties agreed, the continuous motion of the Chariot Race attracted the public's attention but was momentarily eclipsed by a flash of text above, which then faded into darkness and lost its attention value until the next flashed message yanked eyes up again for the few seconds necessary to read it. The discussion in this case made clear that advertisers knew something about the nature of vision. By holding the gaze with visual motion, they knowingly substituted the on-off cycle of the sign's operation for the eye's own naturally occurring saccadic movements. Leading the eye through ocular cycles of adaptation, the flasher mechanism accommodated a physiological need in order to exploit it for commercial ends.[103] To the extent that Times Square had become an arena for reading such illuminated commercial speech, the flash and the interval of darkness were its material integers—solid and void—the building blocks of a new kind of modern space.

In Foucault's discussion of Borges's encyclopedia, the impossible adjacency of irreconcilable elements was possible only "in the non-place of language." Although language can represent such elements, "it can do so only in an unthinkable space. … Absurdity destroys the *and* of the enumeration by making impossible the *in* where the things enumerated" would coexist. Borges, Foucault claimed, "does away with the *site*" of incommensurate appositions, "the mute ground upon which it is possible for entities to be juxtaposed." Borges presented instead a "heterotopia"—the unsettling inverse of a utopia—which Foucault defined as the conjoining of "fragments of a large number of possible orders," each standing apart yet occupying the same space. Heterotopias defy syntax because they hold together things that do not "hold together"; they "dessicate speech, stop words in their tracks, contest the very possibility of grammar at its source."[104]

Foucault might never have visited Times Square, a habitable space that nonetheless contained impossible textual and pictorial adjacencies that stopped words in their tracks. A collection of incongruous pronouncements combined within an urban volume, it created a place that cohered around and because of its feverish fragments. Electricity translated linguistic particles into pulses of light, rich with signification but without a grammar to organize them. With its combinatory structure and fleeting, fluid references, Times Square submerged proximate meaning within what turns out to be a larger discourse *about* meaning. Times Square could not be "read" in any conventional sense, yet it continually invited, indeed, insisted on reading. As the number and complexity of electric signs in Times Square increased, its surfeit of light was supplemented by a surfeit of indecipherable signification. As Chesterton suggested, this new sort of place put extraordinary stress on anyone unprepared to learn a new way to navigate it. Short of the illiteracy he recommended as a means to avert mental meltdown while straining for sense, emotional survival in Times Square required some new perceptual skills.

SPEAKING WALLS

Men, aeroplanes, motor-cars, birds, etc., can be made to move as in a cinematograph, so that a wonderful variety of effects can be obtained.
Eric Norman Simons, 1926[105]

The two great machines of desire in the first half of the twentieth century were the movies and the electric advertising sign, metonymized respectively as Hollywood and Broadway. The new machines of desire created a society in love with representations of desire and in love with representation itself.
William Brevda, 1996[106]

In contrast with other forms of urban reading, whether scanning street signs or combing lines of newsprint, reading became directionless in Times Square. Tempted by color and motion, the eye promiscuously scanned back and forth and up and down, lured and lingering and then leaping from one flash to another to follow wriggling words and cartoon capers, even writing afterimages onto an otherwise unbranded night sky. With each sign touting another special effect, the eye was towed along the square's speaking walls, rehearsing in each ocular orbit a tension between the surrounding scrim of sign faces and resonant pockets of textual, gestural, and pictorial intensity. With motion and electric light the key elements, several models of perception might have recommended themselves to visitors to guide them through the novel experience that was Times Square.

At the most basic level, seeing signs in Times Square might be compared with seeing in an arcade one shop after another, each displaying unlike arrays of goods. Or the continuous scanning of Times Square's night scenery might suggest how a broad landscape is seen, in undulating sweeps interposed with points of visual interest. The news zipper at the Times Tower in fact was said to work on a "panoramic principle."[107] The term surely alluded to several nineteenth-century entertainments wherein viewers watched painted scenes unrolled before them or perambulated a large round room, called a cyclorama, which was painted with picturesque prospects or historical scenes. But the phrase evokes as well "panoramic perception," the German cultural historian Wolfgang Schivelbusch's term for the sort of seeing that arose from nineteenth-century train travel. In this perceptual mode, travelers looking through the windows of moving trains learned to see and comprehend a rushing world. Schivelbusch described the eye's ability to fixate and find a focus in the foreground, which was otherwise blurred by the speeding train: a detail crystallized and then instantly yielded to another passing feature. This novel kind of noticing was tantamount to what Benjamin Gastineau, a French travel writer to whom Schivelbusch referred, called a "synthetic philosophy of the glance" and Schivelbusch summarized as "the tendency to see the discrete indiscriminately," which was characteristic of a new "evanescent reality."[108] More contemporaneously, seeing in Times Square might be likened to what the film theorist Noël Burch described as the topographical vision of early cinemagoers, who "gather signs from all corners of the screen in their quasi-simultaneity, often without very clear or distinctive indices immediately appearing to hierarchise them."[109]

Perception at Times Square shared traits with these sorts of seeing, but, taken as a whole, it was unique. Because the square was a delineated volume, perception was not structured sequentially like the serial scenes Walter Benjamin found in Parisian arcades.[110] Instead, like rail travel, it involved an indiscriminate focus on fleeting details, the licentiousness of which was a good deal of its fascination. The lack of visual hierarchy was readily manifested in the nonhierarchical lists made by visitors to Times Square and, not coincidentally, Gastineau himself, who recounted scenes seen "in quick succession" from the train, "all visions that disappear as soon as they are seen."[111] But in rail travel this manner of seeing entailed directional movement, parallax, and a horizon that structured the visual field into conventions of foreground, middle ground, and background. Most important, it was premised on a view framed through glass. Times Square has little of such elements. Moreover, with words so prominent a feature, Times Square might be better understood as a moving textscape rather than a landscape one travels through. Like the early cinema Burch referred to, Times Square was multifocal, a tableau of tableaux, a single space of overlapping and shifting scenes. But however much New Yorkers and tourists themselves might have compared Times Square to a free show, it cannot be collapsed into a form of cinematic experience.

Seeing was embodied in Times Square signage, and the seeing was unlike the work of eyes and brain in watching movies, which was described by film theorist Mary Ann Doane as a "despatialized" perceptual experience.[112] In Times Square, one shifted weight to maintain balance even while standing still. One was always aware of the crowd, even when it wasn't pressing. Seeing signs and shuffling along with others were congruent at Times Square; they fused, with signs casting color and pattern over the shambling crowd. As Marshall Berman described it, "In the Square the mix is insistently *there*, it's on the street, it's in your face. When you are in the mix, under the Square's spectacular light, ego-boundaries liquefy, identities get slippery."[113] Times Square generated a genuinely new experience of space comprised of the signs' virtual motion, the crowd's omnidirectional ambling, and the consequent effortful experience of reading what's on the move while on the move. Moreover, Times Square's "electric scream" filled the senses and left no latitude for reflection. As Walter Benjamin said in regard to the booming, building-sized billboard, it "abolishes the space where contemplation moved and all but hits us between the eyes," similar to some cinematic effects.[114] The loss of reflective distance is not simply thoroughgoing at Times Square, however. It is the very subject matter of the place, a celebration of indiscrimination.

Further, Times Square obliterated perspective in a manner specific to its historical formation, by subverting the customary coordinates of spatial and linguistic apprehension. In terms of speech, flashers modulated words to the degree that they attained the condition of pictures, thereby collapsing two otherwise distinct classes of visual object into the same cognitive register and aimed at the same aesthetic goal.[115] Flashed words undercut the stability of their signification, abdicating the syntagmatic dimension of their linguistic functioning and thereby renouncing literal meaning. In spatial terms, the middle ground in Times Square dissolved at dusk. The signs loomed large but, owing to early-twentieth century lighting technology, they offered no improvement in detail. The result was a sense of nearness without proximity, an in-your-face intimacy with a cosmic scale, a trait easy to see in the size-related superlatives commentators so often employed. In addition, the triviality of the commodity advertised and its mammoth promotion produced a stark representational foreshortening, which was precisely what writers such as Rupert Brooke were trying to get at with phrases like "celestial underwear," an unnatural joining of opposites. Thus the particular character of Times Square's spectacle rested on an absent middle ground that would otherwise have mediated the shift in visual and cognitive scale between the private act of reading and the alien, autonomous behavior of gigantic pictorialized text. This absent middle ground is the unsteady foundation that Times Square's novel perceptual architecture rested on.

Part of the excitement of urban experience has long been the rearrangement of perspective. The English Romantic poet William Wordsworth, for example, was awed by the random juxtapositions and lurching encounters while he strolled in early nineteenth-century London, and was moved to compile a list of the types of people he came across.[116] Nearly a century later, walking in a built up and gridded city such as New York would have yielded a staccato rhythm of close-in walls perforated at intervals by vistas nominally oriented toward the cardinal points, as if an intervening architecture had segmented the surrounding landscape into its integers. At Times Square, Broadway's disruption of Manhattan's grid created an opening for something else. There the gathering of signs displaced the city's physical order—founded on a language of geometry, durable materials such as brick and stone, and a human scale and orientation—with a new media order. While the old architectural order with its hierarchy of bottom, middle, and top was conditioned by no less an authority than gravity and the strength of materials and articulated with forms that could be traced to antiquity, the new media order obliterated the dimensional references of, say, doors and windows with enormous signs alive with moving text and moving pictures, animate building materials of a new kind of urban space.

SELLING CITIZENS

"Publicity" is the reigning philosophy, the magic conjuring word.
The extent to which you employ it is the mark of your success.
The New Republic, 1920[117]

A search for meaning at Times Square might have been ill-advised, but its electric speech nonetheless expressed something resonant about the American character. The animated signs' short, repeated sequences demanded attention that, despite advertisers' hopes, asked little of memory, a formula appropriate for a nation still young and focused only on what came next. Trade and popular journals breathlessly reported the latest, biggest, brightest, and flashiest sign, effectively challenging yet another sign to supersede it. As a 1911 review phrased it, new electrical displays continually replaced one another so that "the interest in this class of illumination and publicity never grows stale."[118] In 1932 the journalist Mildred Adams grudgingly granted the vibrant verities of Times Square: its signs were "conglomerate and cock-sure, hard-edged, blatant, young enough to turn a full blaze of light into every corner," and they challenged the "ordered and experienced, worldly wise" Parisian manner, which spotlighted only what was tasteful about French society.[119] Indeed, Parisians grew uncomfortable with the thought that New York was becoming the brighter City of Light and were careful to distinguish between their manner of illumination, which was "intellectual rather than electric," from the American version, which was of a more disruptive and grasping nature.[120]

In this way, Times Square represented the untutored energy that had ushered America onto the global stage. Its random exuberance recapitulated the unplanned boisterousness of American business, certified the commercial premise of the American city, and confirmed that the frenzied sales pitch was a national vernacular (figure 5.19). Its sharp-elbowed semantic anarchy was the native patter of free enterprise itself. In Times Square, as much as in the chain stores just then starting to flourish, Americans imbibed the vocabulary and anarchic syntax of capitalist consumption. The signs performed a kind of illuminated discourse on the formation of the commodity. To borrow media historian Michael Schudson's term for advertising in general, Times Square was the urban expression of a "capitalist realism," a picture gallery wherein an economic system represented itself to its

5.19
"What Makes the 'Great White Way' White: So many electric signs they are in each other's way." Brown Bros., photographer. Times Square is held up as a portrait of the competitive practices of American business. "New Wrinkles in Electric Signs," *Literary Digest* 71, no. 2 (Oct. 8, 1921): 24.

imagined beneficiaries.¹²¹ Times Square was in this sense a built pedagogy of American consumer culture, a privately funded self-representation of American national character, reimagined as urban entertainment.

This view offers a new angle for interpreting Walter Benjamin's enigmatic closing statement in "This Space for Rent." Benjamin closed with a question regarding his claim that reflective criticism was no longer possible in a media-soaked world: "What, in the end, makes advertisements so superior to criticism? Not what the moving red neon sign says—but the fiery pool reflecting it in the asphalt." Benjamin's' enigmatic "fiery pool" has been variously understood as an actual reflection in a puddle—a common trope among writers on sky-signs—or an overspilling of vivid color as an autonomous agent of modernity, and more.¹²² But his query specifically concerned the power of persuasion and where it was located. Insofar as Benjamin refers throughout this passage to vision—eyes turned upward to gigantic signs that in turn bombard eyes with commercial messages, images hurled at eyes by film, formerly dry eyes that begin to water at the cinema—the "fiery pool" might well be the collective vision of the crowd, standing on the street and gazing upward at the signaling signs that, in turn, stir a fiery passion to consume. The power of the advertisements at Times Square, then, lies in the commandeered collective attention of a citizenry by now blind to the measured instruction of reflective criticism.¹²³

A LIQUID ARCHITECTURE

I think the huge social and industrial process of America will win in this conflict, and at last capture Niagara altogether. And then what use will it make of its prey?
H. G. Wells, 1906¹²⁴

To be scenically efficient, the cataracts should be visible also at night. ... Just as the cataracts, as a daylight spectacle, have no counterpart in the world, so there would be a matchless riot of electricity here when the shadows have fallen on the earth.
Edward T. Williams, 1916¹²⁵

The image of a small sea of upward-turned eyes staring in rapt wonder in willing and even pleasurable surrender to intense sensory stimulation points toward the sublime, wherein the limits of the mind to process stimuli can, in the right circumstances, become an aesthetic attraction in its own right. In Edmund Burke's eighteenth-century formulation, the sublime emerged from a kind of paralysis—

"astonishment," in his terminology—at the encounter with terror or mind-bending scale. It is a psychic state "so entirely filled with its object, that it cannot entertain any other, nor by consequence reason on that object which employs it."[126] Volition halts, reason ceases; mental activity stalls and submits to the perception of an overwhelming external object. Many accounts of Times Square described precisely such a sensibility, centered on a state of speechless gaping before the visual violence of enormous fevered signs.

The excited rhetoric regarding Times Square, calling attention to sudden shifts in scale, powerful forces, the arrested movement of spectators, and a sense of transporting stupefaction, resonates with language previously reserved for the sublime experience attending visits to natural wonders, which loomed large in European imaginings of the New World. In the early twentieth century, this sense of wonder was already finding its way to manmade objects such as New York's skyline, Henry Ford's River Rouge plant, and, certainly, Times Square. At the time, Niagara Falls and Times Square often topped the itineraries of American tourists to New York State and foreign visitors to America. The two were as far from each other as they could be while remaining in the same state and about as different as two settings could get. Yet the destinations were related in crowning two poles of touristic destinations: one a superlative work of nature, the other a matchless human artifact. Discussions of both sites often employed metaphors of life, energy, and powerful currents, whether of water or electricity at Niagara or of electricity, crowds, or capital at Times Square. Both inspired a sense of awe proceeding from overwhelmed senses that tipped tourists toward feelings of the sublime. Both were said to be ravishing, leaving one disoriented and exhausted and, often, unable to speak.

With the sense of self dissolved before a grand spectacle, both offered moral lessons too, if opposing ones. The sight and sound of the falls were said to ravish even the most armored ego, giving rise to a sense of humility at mankind's transience and insignificance before the works of nature and of God. As one observer put it, "at Niagara, Nature does the talking," while individuals fall silent.[127] The monumental flashing signs and the flowing crowd of the city likewise loosened strictures of civilization, quite a few observed, but only to open the field for clouded judgment and questionable behavior. This was the perilous side of failed memory and forgotten history before the dancing lights of advertising come-ons. As one admonitory text put it in relation to Chicago's entertainment district, "all is levity and enjoyment. It is a living in the present, a forgetfulness of the past, a shutting of the eyes to the terrors of the unborn future." This was likewise one of Times Square's perils. As many noted, an innocent out-of-towner's moral upbringing was all too readily blanched by "the glare of the 'Great White Way.'"[128]

The two sites also bookended developmental narratives, again running in opposite directions. Niagara signified to most visitors early in the nineteenth century a nature awesome, indifferent, and prior to mankind. Over time, Niagara succumbed not only to tourism pressures but, more specifically, to a technological regime of electricity. Starting in 1881, the falls were webbed with wires that within a short time would feed electricity into a local, then a regional, and finally a national power grid. It was cut through by tunnels, diverted by canals, and overtopped by walkways and bridges, reconfigured to serve the power needs of expanding cities as well as the leisured gambols of curious travelers. Niagara was, in the terms of the time, "harnessed" for power and commercialized for tourism. People visited it as much to see the power plants as the falls themselves. For his part, H. G. Wells marveled that a nation capable of exploiting Niagara would then turn all the resulting energy toward "stamping out aluminum 'fancy' ware, and illuminating night advertisements for drug shops and music halls." It proved to him the commercial nature of the American character.[129]

Visitors to the 1901 Pan-American Exposition in Buffalo did not even need to go to the falls to appreciate their power. The fair's Court of Fountains thundered with rushing waters; the Tower of Electricity, rising nearly 400 feet and covered with panels "intended to suggest the water as it curves over the crest at Niagara," translated the cataract into a skyscraper.[130] Fountains, colored tiles, statuary, and flashing lights: all were homage to the yoking of the Niagara River for the production of electricity. Many speculated that the time would soon come when the power of the falls would be available to distant cities, "even as far from Niagara as New York."[131] Perhaps the most vivid demonstration of the complete transformation of Niagara Falls came in 1907, when a battery of powerful searchlights transmuted its pounding cascades into currents of electricity that powered batteries of lamps to throw on the falls a play of flickering, ephemeral effects to delight sightseers.[132]

Times Square, in contrast, began just as Niagara Falls ended its run as a sublime work of nature. It sprang from a confluence of electricity and business interests that blossomed into a place so completely commercial and complexly contrived that its artificial atmosphere was soon taken as a world of its own, a kind of second nature. While the power of the falls was harnessed to generate electricity, electrical energy was unbridled at Times Square in torrents and shimmering sprays of light. Whereas the falls were the authoritative voice of nature expressing transcendent themes, Times Square tittered on with gags about commercial goods. Tourists liked them both.

The connection between the two sites was recognized almost as soon as Times Square started to glow. Admittedly partisan in acclaiming Times Square, O. J. Gude made the connection explicit: "It has become an accepted truism in Europe that the three great memories the traveler brings back with him from America are the wonderful, awe-inspiring Niagara Falls, the gigantic skyscrapers of New York City and the electric advertising signs of the 'Great White Way.'"[133] Soon after, the Russian émigré essayist and journalist Simeon Strunsky, taking note of the "monster electric sign-boards" that "gleam and flash and revolve and confound the eye and senses," christened Times Square an "electric Niagara,"[134] while another writer imagined a bemused Martian seeing in Times Square "a Niagara of electricity, which pours a flood of its fire over the fronts of buildings."[135] The analogy returned episodically and was most memorably embodied by a 120-foot-wide waterfall that flooded the square with light. Stretching from 44th to 45th Streets, the display was made in 1948 by Douglas Leigh for the Bond clothing store (figure 5.20). Marshall Berman recalled the hypnotic trance he felt standing before it, as if Niagara had not only surrendered its energy to Times Square but had also handed off its power to inspire awe.[136]

5.20
The world's greatest artificial waterfall, recalling Niagara Falls both as form and as a source of electricity. *Largest Spectacular Sign in the World*, Acacia Card Co., New York. Tichnor Brothers Postcard Collection, Boston Public Library.

SHARING PROGRESS

The Great White Way

Then, glory of glories, the town put in a White Way.
Sinclair Lewis, 1920[137]

Every "Main Street" is a miniature Broadway.
***The Literary Digest*, 1921**[138]

The reputation of Broadway and Times Square sparked a nationwide "white way" movement in the early decades of the twentieth century, cheered on by a host of advocates (figure 5.21).[139] Manufacturers of electrical equipment specialized in "white way" systems, usually consisting of incandescent bulbs encased in translucent globes and hung in clusters from decorative lamp standards. Local utilities, for their part, sought to develop nighttime load to exploit the generating capacity they had built up to meet heavy daytime loads from offices and factories. Merchant groups and civic boosters believed that brighter lighting attracted more customers, or at least prevented existing customers from decamping to lights newly installed along a competing business street. The economic rationale appeared self-evident: brighter lighting attracted people, higher levels of pedestrian and vehicular traffic augmented commerce and inflated property values, and high land prices intimated economic desirability, which in turn was a firm foundation for civic pride. According to its advocates, white way lighting was as an investment in the future rather than a present-day expense. Even if specific measurements were hard to come by, it initiated a self-fulfilling progressive trajectory: it "gives to a city an air of progressiveness and prosperity. 'Nothing succeeds like success.' To appear prosperous is the first step to being prosperous," according to a representative claim.[140] The effects—a "psychological impression of thrift and progress"—might be subjective but the profits, advocates insisted, were real.[141] By 1912, and counting only new installations using the recently introduced tungsten lamps, there were at least 220 "Gay White Ways" within the United States alone.[142] More than a few cities boasted that their white ways made them "the best lighted city in America."

In the narrow sense, white way lighting involved street lamps installed by a municipality or merchant association along a major business route. But in the popular imagination it was coupled with advertisements. Times Square, with private signs outshining public street lights, was the touchstone.[143] The electric signs in Auburn, New York, for example, were "monuments to the progressiveness of the merchants of the city," crowed an advertising manager, and were erected in the same spirit as those in Manhattan, some 250 miles downstate.[144] Los Angeles, an urban lighting pioneer in its own right, outshone New York since, its boosters argued, its White Way was brighter on a per capita basis.[145] In a 1915 report on new white ways being built, a business journal noted: "From one end of the country to the other people have heard of, if they have not seen, New York's 'Great White Way.' It was the first of what is now a great host."[146] By 1936, H. L. Mencken could claim that white ways had supplanted allées as evidence that yearning for material prosperity had trumped respect for civic decorum: "Every American town of any airs has a Great White Way; … in the Era of Optimism, rows of fine shade-trees were cut down to make room for them."[147] Times Square had spawned "the 'commercial aesthetic'—bright carnivalesque signs of color, light, and glass designed to thrill, excite, and awe onlookers while convincing them to buy," which went on to influence a large swath of American culture. It "was setting [the] pace for the nation."[148] More than a few cities, in other words, believed that commercial and, by extension, civic progress involved building for themselves a zone of illuminated commercial speech.

5.21

Every city or town of any aspiration acquired a Great White Way in the early twentieth century. Captioned "A real white way—Church Street, New Haven, Conn.," in John Allen Corcoran, "The City Light and Beautiful," *American City* 7, no. 1 (July 1912): 47.

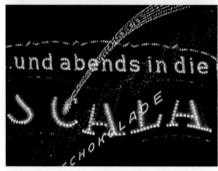

5.22
A sequence of large illuminated advertising signs blend and overlap to evoke a new form of urban reading. Film stills of collage sequence of spectaculars in *Berlin: Symphony of a Great City*, directed by Walter Ruttmann (1927).

Times Square Abroad

This glittering trail along upper Broadway,
the "Great White Way,"
is celebrated all over the world.
New York Herald, Sept. 22, 1906[149]

In 1904, Stephen Chalmers, a New York booster, acknowledged that Broadway was not entirely unlike other main streets around the world, brimming with "the gayety that always accompanies artificial light." But it was the greatest of all, he continued, it "outshines Piccadilly Circus and its other rivals in Paris, Port Said, and Vienna in that it has the essential features of every one of these," superadded to which was an American "native hue of resolution. ... We will grant for the one-millionth time that Broadway is the greatest thoroughfare in the world, for you would smile at any one who said it was not."[150] In following years, Broadway and Times Square suffered no lack of backers. By 1927 British writer Stephen Graham had seen the imitators but maintained that New York's Broadway was "the mother of Broadways all over the world, mother of the lights of Piccadilly Circus and of the Place Pigalle and Teatralny Ploshtchad. The Great White Way is the greater white way" (figure 5.22).[151] Calling it "luminous epilepsy, incandescent hypnotism" when he visited in 1930, the banker, writer, and vintner Philippe de Rothschild estimated Times Square's emotional value as being greater than that of "a hundred Eiffel towers, a thousand Rue Pigalle." Considering by way of contrast the protracted pace of sidereal motion, he quipped: "Pity the sky with nothing but stars."[152] While Times Square did not invent illuminated commercial speech, it nonetheless became its symbolic home.

Today, Times Square is no longer the brightest or the most extensive zone of illuminated commercial speech. Virtually every large city on Earth features a district dominated by flaming walls of animated winks, dances, and underwear. Around the world, from Asakusa Rokku in Tokyo to Avenida Corrientes in Buenos Aires, theater districts have electrified and embellished their lighting (figure 5.23). These are joined by what are now called "urban entertainment zones," places brimming with distractions, shopping, restaurants, and street life and bathed in the light of overscaled flashing signs. One example, Bukit Bintang, in Kuala Lumpur, even features a shopping mall, one of the world's largest, named for Times Square. Larger districts, such as Fremont Street and the strip in Las Vegas or the Pudong quarter of Shanghai, with hectares of glinting spectaculars, are easily understood as metastasized versions of Times Square. There is even an intergalactic Times Square, the Uscru Entertainment District, which has been a backdrop in several of the *Star Wars* films but remains fictional for the moment, as far as we know.

Plural Times Squares, in New York and around the world, are the ubiquitous urban beacons that proclaim the global contract between capital and urban identity. All these places have distinct histories and diverse geometries, but the experience of visiting them is surprisingly similar. The people there, the goods for sale, the languages spoken, and the smells wafting about—all are different enough to be easily distinguishable. But all are overlooked by the same cast of coruscating divinities of illuminated commercial speech. Times Square has crystallized into a reference point for global culture, a district of unabashed luminous business promotion and a leading feature of the global city.

5.23
The fiery pool of reflected light now appears in cities around the world. *Kabukicho in the Rain*, B. Lucava, CC BY-NC-ND 2.0.

6
GROPING IN THE DARK

Our human nature is profoundly phototropic.
György Kepes, 1965[1]

INTRODUCTION

New dim-out regulations taken from the Independent: ...
This is very dry reading, but in after years it will remind us
of the weird, ghostly streets, the phantom headlights,
the mad scramble to find black material to cover our windows
and to some of us, the one blackout room where
we spent our evenings.
Mable R. Gerken, 1943[2]

Government-mandated blackouts precipitated a crisis in the optical and spatial consciousness of the American public just before and during the first years of World War II. In an effort to foil potential aerial bombardment, citizens were asked to turn off their lights and so break an otherwise unqualified promise of modernization: ubiquitous illumination (figure 6.1). After decades of constantly increasing levels of artificial light, blackouts challenged not only nighttime visibility but spatial perception more generally. Americans discussed ways to adjust to dimmer surroundings, to infer spatial information from nonvisual senses, and to familiarize themselves with nightscapes based on reflective rather than geometric properties of surfaces. They learned to inhabit scotopic space, that is, spatial relations as inferred under conditions of low lighting, in contrast to the sense of space that obtains under photopic, or daylight, levels. Although the blackouts lasted only a few years in the United States, they reveal the general outline, though in negative terms, of how the nation's visual environment was understood following decades of abundant electric light.

Encouraged by the federal government, and nominally ordered by branches of the military, blackouts were always implemented state by state and town by town by a typically volunteer force of local air raid wardens, usually with little training. The complete extent of blackouts, the towns and cities blacked out, the number of citizens who sat in darkness for varying periods of time, the number of blackouts that were just for practice or were imposed out of fear of imminent attack—these quantities are not known with certainty. Nonetheless, the volume of statements issued from the widest range of sources, spanning private businesses, educational and research institutions, and all levels of government, reveals the degree to which blackouts became a significant, if short-lived, part of national imagination during wartime. In the United States, where aerial bombing was anticipated but never experienced, the discourse of scotopic space demonstrates the degree to which electric lighting had been naturalized as a normal dimension of lived space. It constitutes a collective portrait of a new type of perceptual space characterized chiefly by the sudden and surprising loss of electric light.

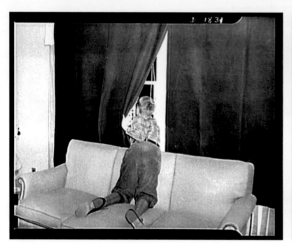 

6.1

Images disseminated by the U.S. Office of War Information and captioned "Looking out the window is fruitless also because there is nothing one can see during a blackout" and "Air Raid Protection—what NOT to do when alarm sounds. DON'T look out the windows." Farm Security Administration/Office of War Information Collection, Library of Congress, Prints and Photographs Division (http://www.loc.gov/pictures/resource/fsa.8b01242, and https://www.loc.gov/item/2017699562).

AN EVER BRIGHTER FUTURE

During 1928 the electrical industry continued that
steady growth which has characterized it for
the past twenty years.
E. M. Herr, 1929[3]

As much as the 1929 Electric Light's Golden Jubilee celebrations were an industry stunt—organized by the General Electric Company and executed to a large degree by Edward Bernays, a pioneer of modern public relations—they also registered a basic truth about the meaning of electric light, in addition to its remarkable proliferation. Even in the darkest days of the Great Depression (the jubilee culminated one week before the 1929 stock market crash), electric light seemed a lone promise of continued advancement: bulbs were getting cheaper, brighter, and more durable; the cost of generating electricity was dropping steadily; and regional grids were being linked into a national network. Investigations into monopolistic behavior on the part of the electrical industry did not seem to mute the popular opinion that it was an engine of economic growth and in any case was only responding to competitive pressures from European consortiums.[4]

Areas with low population densities had been neglected for more promising markets by the industry, but the 1936 Rural Electrification Act extended government loans and new technical standards and services to help spread the franchise of electric light and electric power to farms and other rural areas (figure 6.2). For its part, the Illuminating Engineering Society (IES) recommended ever higher levels of illumination across the range of all building types, with most customers following the drift toward more and brighter light. In 1938 a new sort of lamp that had been researched for more than 100 years was introduced commercially: fluorescent light. The following year it was one of the marvels at the New York World's Fair, along with FM radio, television, and working robots. Sales of fluorescent lamps increased from 200,000 in 1938 to nearly 34 million by 1942, and they were predicted to replace incandescent sources within a short time, allowing homes and businesses to be illuminated even more brightly and at lower cost.[5] Electric light was a luxury in the 1880s, and a new installation was a newsworthy event. But fifty years later, with at least several bulbs burning even in modest homes, electric light made manifest an egalitarian spirit believed by many to be uniquely American.[6]

A nostalgia for darkness was the surest sign that electric lighting had become naturalized by the 1920s. Even earlier, a blacked-out London in World War I was said by some to offer *un nouveau frisson*, a new way to see the city, at least initially.[7] A decade later, the *New York Times* writer Mildred Adams argued that New York had become less interesting in visual terms since it was electrified. A visitor drawn

to dazzling spots such as Times Square would unwittingly "miss some of the most picturesque and dramatic aspects of the city. He does not know New York's possibilities who has not seen Wall Street when its empty canyon is drenched in night." It was less a matter of urban variety than of achieving richer sensory impressions of the city: in the dark, buildings "mingle with the texture of the night into a blackness so thick that one can almost touch it." Sounds were sudden and sharper in the dark: "A laugh is as startling as a gunshot."[8] In this and similar statements, electric light had eclipsed other perceptual inputs and aggrandized vision. In doing so it had emptied nighttime experiences of broad sensory potential. Clearly, defense of the dark was premised on the reliable presence of bright light everywhere else.

Decades of visual dilation by electric light seemed only to reaffirm sight's long-standing status as the primary channel through which knowledge of the world would be gathered. The presumed perceptual supremacy of vision for the widest range of tasks, for the conduct of everyday life, and as a foundation for the perception of space could rest as easily on tungsten filaments as it did on celestial bodies (figure 6.3). Not only did electric light brighten formerly dim spaces, it brought with it a new intensity to visual experience, fostering a sense that people were inhabiting a new affective environment. Electric light did not just improve sight, it stimulated the entire sensorium. According to a 1903 study published in the *American Journal of Psychology*, it produced "certain psychic and physiological changes of so marked a character as to place them in a class by themselves." More than three quarters of the study's nearly 300 subjects said their spirits rose when they walked into a brightly lighted room.[9] This emotional uplift was believed to have an effect on intellectual effort as well, as mentioned earlier in relation to lighting in the workplace. In time, as electric light was incorporated into a wide range of buildings and activities and gradually began to lose its extraordinary character, its variable brightness, ready accessibility, and convenient placement came to be an expected rather than an exceptional condition of the everyday environment.

6.2
Spreading the franchise of electric light nationwide. Lester Beall, *Light—Rural Electrification Administration*, 193–. Library of Congress Online Catalog, Prints and Photographs Division (https://www.loc.gov/item/2010646236).

6.3

A series of advertisements in the Maxfield Parrish style deployed cosmic and mythic symbolism to suggest that night would henceforth be as bright as day. "His Only Rival," 1910, General Electric advertisement, color lithograph (printed by Ketterlinus, Philadelphia). Reproduced with permission from the Archives and Research Center, Museum of Innovation and Science, Schenectady, New York.

A BLACKOUT OF PEACE

And now darkness. A new world. Black-out, bombs, slaughter, Nazism. Now the night and the shrieks and barbarism.
William L. Shirer, 1939[10]

Enemy bombs may well be dropping on American cities before the end of 1943.
James M. Landis, 1943[11]

At exactly the same time, however, fears of nighttime bombings were rising around the globe, with mandated blackouts a frequent response to growing threats of hostilities. Combatants had experimented with aerial bombing as far back as the nineteenth century, with soldiers tossing explosives from gondolas suspended beneath hot-air balloons. Efforts accelerated in World War I, with Germany initiating the practice in the summer of 1914, dropping bombs over Belgium from zeppelins, and England following suit that autumn with bombing runs over Düsseldorf and Cologne in the west of Germany. In response, many cities began to require their residents to black out their homes and businesses in anticipation of air raids.[12] Militarily, the blackouts were said to be effective, leading enemy aircraft, which could be difficult to maneuver, to lose their way, sometimes diverting 90 percent of their bombs from their intended targets and so contributing little to the strategic execution of the war.[13]

Nonetheless, the idea of bombs raining from the sky on civilian populations was horrifying to virtually everyone as World War I advanced. Many, doubting that it was possible to defend against aerial attack, began to consider the inevitability of heavy losses in urban centers.[14] After the war, in 1923, a number of nations met in The Hague to draft rules for aerial warfare. The League of Nations would continue to revisit the issue, culminating in a 1937 resolution condemning aerial bombardment of civilian sites.[15] Alarmed, governments drew up plans for ways to defend against the practice. But aircraft capacities were advancing rapidly, and the prospect of putting up adequate defenses to patrol thousands upon thousands of square miles of airspace was daunting. Experts continued to raise doubts that any defense was possible. As the soon to be British prime minister Stanley Baldwin said in a 1932 speech to Parliament, "The bomber will always get through."[16] The stakes seemed only to get higher as the decade wore on. In 1935 Germany began to rearm, Italy attacked Ethiopia, and Franklin Delano Roosevelt began to voice his concerns about being drawn into war.[17] That same year, blackouts occurred in Gibraltar and Istanbul. Madrid imposed blackouts in 1936 after rebel bombings. In 1937 military factions in governments in Germany and Japan called for blackouts

6.4
Hostile enemy airplanes might be just over the horizon. James M. Landis, "Get Ready to Be Bombed," *American Magazine* 135, no. 6 (June 1943): 30–31. Illustration by Fred Ludekens.

6.5
If Chicago Were Bombed. Preparations for aerial bombardment were taken seriously, even for cities 1,000 miles inland. "Is Your Water-Works Plant Prepared?," *American City*, August 1942, 9. Copyright 2017, Penton Media, 131085:0917SH.

in Berlin and Tokyo, fearing retaliatory air raids; Guernica was attacked from the air, with calamitous results that shocked millions around the world; W. H. Auden and Christopher Isherwood traveled to China to write about, among other things, aerial attacks on Chinese towns. Other writers, such as H. G. Wells in his 1933 story, *The Shape of Things to Come*, envisioned bombs and "heat-rays" showering devastation on urban centers. The decade swelled with a new awareness of civilian vulnerability to death from the sky.[18] By 1939, and especially after the start of World War II, blackouts had become common in German cities; in Poland, where Germany began bombing towns in 1939; in London, Paris, and Rome; and in China and Japan. To many, the world seemed suddenly to have been thrown back to pre-electric, even preindustrial times. Britain, for example, was said to have acquired a "medieval flavor" in 1939, with December 25 of that year proclaimed the "first blacked-out Christmas in history." Later accounts spoke of "the cave-man's existence to which we were now condemned" in the blackout.[19]

Although separated from Europe and Asia by thousands of miles of ocean, Americans grew edgy (figure 6.4). The capacity of long-range aircraft was as uncertain as the malevolence of militant foreign regimes. Anxieties waxed in anticipation of what Winston Churchill would term "a new kind of war," one that made the home front a target, an active theater of operations as well the foundry for soldiers and matériel.[20] Isolationists warned against American military involvement overseas, but some experts described the potential of long-range aerial raids on the United States to unleash vast destruction and draw the country into conflict regardless of its hesitation. The American government began requisitioning military equipment and set up air raid protocols, urban service facilities made emergency plans and took protective measures, and municipalities began to hold blackouts to practice concealing their populations from airborne enemies (figure 6.5).[21] The first American blackout took place on the evening of May 16, 1938, in Farmingdale, on Long Island, with cities and towns across the nation, including Honolulu, blacking out the following year (figure 6.6).[22]

To prepare for the approaching tide of darkness, Americans were urged to imagine what was already happening in Europe: "If at 9:30 o'clock it got darker than you had ever seen Manhattan and every light on the Great White Way went out and every movie and every single show closed, and cars crept along dark streets in second gear with only vague blue lights showing on the ground, you would have a glimmer of an idea of what a London blackout is like."[23] Soon, Edward R. Murrow began describing to Americans the eeriness of life in the London blackout. He registered as well the rearrangement of sensory inputs in a darkened city, telling his listeners in a radio broadcast on August 24, 1940:

6.6

Immediately after Pearl Harbor, American cities begin blackout drills. *Los Angeles Gropes in Its First Black-Out. Los Angeles Times*, Dec. 11, 1941, 1. Copyright © 1941, *Los Angeles Times*. Reprinted with permission.

6.7

Blackout in 1942, Keene NH. Before and after views on March 12, 1942. Towns large and small across the United States held blackout drills. About 14,000 people lived in Keene, New Hampshire, at the time of the blackout. Courtesy of Keene Public Library, Keene, New Hampshire.

I'll just let you listen to the traffic and the sound of the siren for a moment. … I'll just ooze down in the darkness here along these steps and see if I can pick up the sound of peoples' feet as they walk along. One of the strangest sounds one can hear in London these days—or rather these dark nights—just the sound of footsteps walking along the streets, like ghosts shod with steel shoes.[24]

The New York World's Fair was blacked out on October 15, 1940, with, one presumes, even the bright new fluorescent lights getting switched off. Indeed, the simple fact that the lights at the fair burned at other times became an eerie reminder of a world at the edge of an abyss: "It seems like a page torn out of an old book when the world was innocent and having a good time was a normal pursuit for human beings. There were no blackouts then to make the many-colored glow of our World's Fair seem infinitely precious."[25]

In retrospect, aerial bombardment was not a serious threat. But at the time the range of airplanes and the size of their payloads were uncertain, and rumors spread of jet-propelled bombs that would become all too real with the V-1 and V-2 rockets later in the war. Blackouts and dimouts had in fact proven effective along the East Coast of the United States, where German submarines had been known to prowl, by greatly diminishing light that could be seen even miles out to sea. Moreover, many Americans were motivated to do as much as possible to help the war effort. Support for blackout drills was understandably robust on the East and West Coasts, but it was also strong in the Midwest, where there was little chance of bombing attacks. About a third of midwesterners believed they should hold blackout drills (figures 6.7, 6.8, and 6.9).[26] Many women, it was reported, felt reassured by taking precautions such as readying their homes for blackouts, feeling their "mental balance" had been restored after the initial shock of hearing war declared.[27] Moreover, stoking fears of enemy attack was one means of overcoming isolationist sentiment in the years leading up to American involvement in the war, and an impetus for cooperation among the widest range of civilians.[28] As hostilities in Europe heightened, Roosevelt began to use his communications with the public, including his fireside chats, to help Americans mentally prepare for war.[29]

Regardless of the likelihood of aerial bombardment of civilian populations, the American context of blackouts is valuable to consider for understanding the degree to which electric lighting had become a premise of modern perception precisely because so much of it was conjectural. Without the pressure of immediate peril, blackouts were free to expand in the cultural imagination during the early years of World War II. Advice for managing life in a darkened world could flourish when blackouts were short-lived and nonthreatening, and when no budgets or life-and-death decisions were present to dampen speculation.

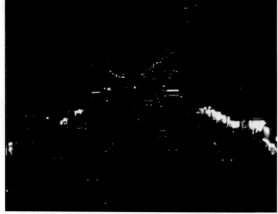

6.8
"Take a look at the bright lights of New York for some time to come. ... No more dazzling lights on Times Square." Before and after shots were a common way to dramatize the severity of the blackout. Film stills, *Blackout in Manhattan* (Dublin issue), *British Pathé*, Nov. 6, 1942.

6.9
Test blackout in Salt Lake City, more than 800 miles from an ocean, on the first anniversary of the bombing of Pearl Harbor. The well-defined volume of Main Street evident by day is rendered by the blackout's intermittent lighting as incongruent planes and floating squares. *Salt Lake City—Main Street Blackout—shot 4*, Dec. 7, 1942. MSS C 400 Salt Lake Tribune Negative Collection, Utah State History. Used by permission, Utah State Historical Society.

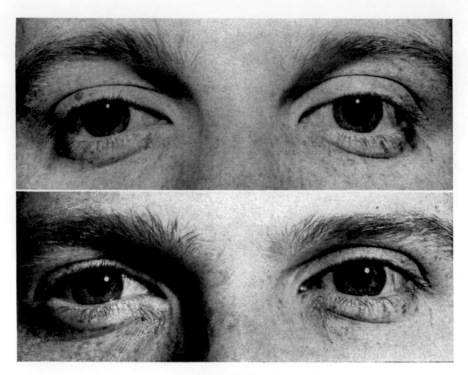

NIGHT EYES

The pupils of the eyes expand (top) to let in more light in darkness, but are normally smaller in daylight.

6.10

"Night Eyes." Popular magazines explained the physiology and the physics of night vision to help people understand the difficulty of seeing at night, for military and civilians alike. "How to See at Night," *Science News-Letter*, June 6, 1942, 358. Reprinted with permission from *Science News*.

SPATIAL INTUITION IN A BLACKOUT

Two thirty a.m.—Almost through our first black-out.
The city is completely darkened. It takes a little getting used to.
You grope around the pitch-black streets and pretty soon
your eyes get used to it.
William Shirer, 1939[30]

The shadow of the war fell still more literally on
New York last night as the dim-out put in force by the
Army the night before was extended to include the
world-famous advertising signs of Times Square. ...
Throngs moved through the comparative gloom of the
canyon that is the square and seemed puzzled.
***New York Times*, 1942**[31]

After decades of rising expectations regarding levels of illumination, the idea of a darkened nation came as a shock, even though blackouts were sporadic and announced ahead of time. Where vision had been improving year to year, seeing suddenly became harder. Numerous publications, issuing from dozens of distinct professions, rushed to address the question of "seeing in the dark." Nearly all began with a sketch of the eye's anatomy and the physiology of vision, and then moved to draw a crucial "distinction between day and night vision in the psychological technique of seeing" (figure 6.10).[32] Seeing involved the mind, not just the eye. Advice, therefore, took the form of numerous suggestions useful in a range of situations. For example, objects can be seen better at night when fixated peripherally rather than directly, although the resulting image cannot be sharply focused. A spread in *Life* explained the physiology of the eye under varying intensities of light and illustrated how "pilots, plane-spotters, coast guards, sailors and sentries on night duty" had learned to see in the dark by looking at objects obliquely so that their retinal image fell on light-sensitive rod-shaped cells outside the fovea, the retina's most visually acute spot.[33] Readers were also advised to recall daytime details to supplement their nighttime vision. Since the retinal pigment visual purple, or rhodopsin—a protein in the retina that is crucial to converting light into electrical signals for the brain to process—loses sensitivity under steady light, keeping the eyes in motion would help too. Momentary exposure to light could undo many minutes of adaptation to the dark—one minute to recover sensitivity for each second of exposure was a common rule of thumb—so those with a critical role to play in blackouts, such as air raid wardens or airplane spotters, were advised to wear dark goggles or a patch over one eye so that at least monocular vision would be ready for the dark. Contrast was as much a factor as absolute brightness, so even a candle or flashlight could diminish one's night vision.[34]

City dwellers were said to be especially at risk, having lived the longest with bright lights at night. Researchers had earlier noted that the configuration of light-sensitive cells in the retina was remarkably adaptable and, they argued, could evolve under continued exposure to artificial light. A 1908 study speculated that urbanites would in time lose their night vision, although they would gain in color discrimination as the eye's cone cells replaced atrophied rod cells.[35] Reviving this claim, the *Washington Post* warned in 1942 that "a large majority of persons will have to learn to see all over again. … When certain muscles of the eye are not used for a considerable length of time they grow flabby and can't do the work for which they were intended. Thus country people, long used to moonlight and the darkness of moonless nights, probably will be of more value in blackouts."[36] Everyone was advised to eat lots of carrots and, if they could stand it, fish oil, as many studies confirmed that vitamin A was essential to good night vision. Untroubled visual perception in the crisply lit world of electric light was swiftly replaced by a dawning awareness of the mechanics and the labor of seeing.

With the apparent spontaneity of sight compromised, magazines and newspapers, along with blackout and civil defense pamphlets, argued that conscious attention had to be redirected toward the other senses. In the dark, individuals could communicate with a system of audible cues, for example; they already recognized sirens that signaled the beginning and end of blackouts. In spatial terms, sound could help determine an object's position, although figuring distance through sound was difficult. Curt Wachtel, a German émigré pharmacologist and a leading expert on chemical warfare (he had tested hundreds of toxic gases for the German government during World War I), urged adults to become familiar with odors, partly in anticipation of poisonous gases and also in the event of household dangers resulting from bombing damage. He cautioned readers to sniff rather than inhale deeply, which decreased olfactory sensitivity, and he advised the development of a standard vocabulary to distinguish and communicate the presence of odors that might signify hazards, including "sour milk, sour cucumber, herring, decayed straw, horseradish, mustard," and so on.[37] The skin, too, had a role to play. Although vision usually dominated spatial apprehension, it was supplemented and, in many versions, preceded by the sense of touch, which, a number of perceptual psychologists argued, was both a prior and a purer form of spatial intuition. In addressing "the problem of space" in 1937, Géza Révész, a Hungarian psychologist who specialized in aesthetic perception, argued that while visual space was superior to haptic space—the sense of space generated primarily by touch—in its scope, detail, and immediacy, it was nevertheless structured on earlier tactile experience. Whereas touch could yield spatial information without sight, a purely visual intuition of space was impossible.[38]

With the other senses summoned, memory was said to be a more active mental faculty in assembling a sightless sense of space. The nonvisual senses relied on "mnemonic traces," wrote the psychologist Walter Blumenfeld in 1937.[39] Memory jointed successive sensations to one another to produce a robust understanding of the surrounding environment. Touch, said Blumenfeld, involved "a stage by stage making structured the world of objects by means of movements," that is, an incremental gathering and ordering of tactile sensations into an operable concept of space. Groping in the dark, a cliché commonly applied to movement during blackouts, describes perfectly such an incremental discovery of space by tactile means. Finding that "the haptic and the optical construction of space is governed by parallel laws," Blumenfeld rejected the conclusion drawn from everyday experience of "one single phenomenal *structure* of space—that of the optical apparatus" and argued instead for a visual apparatus that was fundamentally hybrid and closely related to motion, as well as to touch. Nevertheless, he continued, vision was such a powerful means of apprehending space that even while feeling one's way through the dark, tactile stimuli would be related to a visual schema, with each touch taking its place in a conception of the physical world that was already internalized based on prior optical experience.

The choreography of the senses in pursuit of an operable spatial construct was familiar to the general public through its blackout experience. Commonsense tips that hinged on the calculated coordination of different senses were repeatedly proffered in many disciplines' protocols for blackouts. In the medical literature, for example, contributors noted that many services were already delivered by hand. Experienced fingers could palpate patients in the dark and with additional practice find an upraised vein for an injection. Likewise, filling a syringe in the dark, the writers reported, was simple. Finding equipment in the middle of the night, however, was harder. One doctor advised fellow practitioners to wrap equipment in different kinds of fabric, layer instruments within a medical bag, and place the bag in a specific location so that, when needed, it could be found in the dark.[40] Air raid wardens similarly described what can be termed a procedural apprehension of space. One night in 1943 near Los Angeles, for instance, air raid warden Bill Barbour "rolled himself to a sitting posture on the edge of the bed, and groping in the blackness about his feet, found his socks and shoes. He stood up against the chair on which, retiring last night, he had piled his clothes in canny successive order; reversing the order, he put them on one by one."[41] Tactile and motor memory is in these instances codified in terms of procedures that compensate for the loss of visual information even if, as Blumenfeld held, the operative spatial schema is still visual.[42]

The overall picture that emerges from the blackout literature was that scotopic space, the mental construction of space under conditions of low light, was thought to be the product of "somatic attention," a term describing the conscious awareness of one's sense organs, along with the sensations they derive from the world. Such somatic experience is "inherently plastic"; that is, attention to sense organs, along with the sensations they harvest, alters the meaning—the very content—of bodily experience.[43] The term "somaesthetics" is recent, referring to the disciplined study of the role of the body in relation to aesthetic experience, broadly conceived. But the idea is older. Blumenfeld had already argued that sentient beings innately harbored an interest in "self observation" that often resulted in "the proprioceptive point of view taking precedence for a long time over the exteroceptive"—in other words, conscious attention to the sense organs and the muscles that operate them eclipses the mind's attention to sensory input from external sources.[44] With its perceptual objects spatially localized outside and separate from the eye, vision is the most exteroceptive of senses and, since the effort to see hardly registers in relation to the information received, the visual world can appear self-evident and given. In contrast, perceptions gained through touch, for example, fuse apprehension of the sensing self with sensations from a world beyond the body. The world becomes more subjective this way, when, in addition to gathering sensations, the self senses itself sensing. What Blumenfeld called "the spatial threshold of the skin" was thus a surface on which perceptions of outer and inner worlds overlapped in the apprehension and understanding of space.

6.11
Lights are recoded during the blackout in terms of urgency. Automobile headlights, properly shaded, are necessary at night, whereas the theater marquee at the Astor Theater in Times Square could be safely darkened in a series of dimout and blackout drills in New York. The looming presence of the giant marquee is visible only at the edge of sight. *New York, New York. Times Square during the Wartime Dimout at Night*, Marjory Collins, photographer, September 1942. Library of Congress Online Catalog, Prints and Photographs Division (https://www.loc.gov/item/owi2001011719/PP).

6.12
Reflective properties of materials in an urban setting take on heightened significance in a blackout. *Blackouts* (Washington, DC: U.S. Office of Civilian Defense, August 1941), 5.

6.13
Augmenting the reflectivity of objects in a rural environment. *Blackouts* (Washington, DC: U.S. Office of Civilian Defense, August 1941), 54.

THE CHARACTER OF SCOTOPIC SPACE

Is there really a special kind of vision during a blackout?
How well can we see with it? Is it true that a match
is visible for miles?
Selig Hecht, 1942[45]

Blacked-out urban spaces were often described as fragmented. The fabric of light woven from street lamps, electric signs, radiant shop windows, glowing rectangles of home windows, and waves of automobile headlights came undone in the dark (figure 6.11). Windows were blacked out, shops closed at sundown, street lights went off, drivers stayed home. The luminous form of cities became granular and episodic, patches of light rather than a continuous "river of light," a phrase used in prior decades to characterize electrified streets, especially bright ones like New York's Broadway. Where light once flowed, in a blackout it leaked and seeped, often in geometric forms such as beams, nicks, slits, and slots. With occasional points and gleams difficult to resolve in terms of distance and with other visual signals of spatial

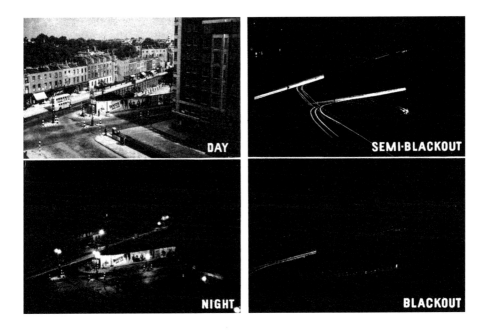

6.14
The image of the city as it changes from photopic
to scotopic conditions. *Blackouts* (Washington, DC:
U.S. Office of Civilian Defense, August 1941), 3.

6.15

"Hazard." Inside the home, new ways of comprehending and navigating the darkened realm of the blackout are needed. "How to Guard against Home Blackout Accidents," *Science News-Letter* 41, no. 6 (Feb. 7, 1942): 84. Reprinted with permission from *Science News*.

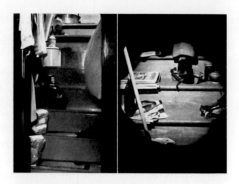

6.16

Building family camaraderie by preparing the house for blackouts together. "Air raid protection—what to do in your home." U.S. Office for Emergency Management, 1941[?]. Farm Security Administration/Office of War Information Collection, Library of Congress, Prints and Photographs Division.

6.17

Home leisure activities can be adjusted to suit the amount of light shed by a single shaded lamp. "Air raid protection—what to do in your home." U.S. Office of War Information, Overseas Picture Division. Farm Security Administration/Office of War Information Collection, Library of Congress, Prints and Photographs Division (http://www.loc.gov/pictures/resource/fsa.8b01226).

character dissolved in the lowered light, the urban night became more two-dimensional than it had been for decades. To put it another way, the perceptual profile of the blacked-out city shifted from being a product of outline, color, distance, rendered volume, detail, and so on, under photopic conditions, to being a product of the reflective properties of surfaces (figures 6.12 and 6.13). Space in turn became understood as more of a specular than a geometric construct; the metric of visual space was given not by solids and voids—buildings and the intervals between them—so much as by relative degrees of reflectivity.

Furthermore, the perceived threat of air raids caused light to be recoded in terms of its urgency or relevance to the larger war effort (figure 6.14). Unnecessary lights were extinguished, useful lights were dimmed, and essential lights were kept indoors by sealing openings through which light might leak. Even a match or a glowing cigarette could be seen from miles away, it was said, so smoking outdoors was banned during blackouts and offenders were fined. Such strictures were quickly internalized, with citizens sometimes taking enforcement into their own hands. One man was sent to the emergency room of a Brooklyn hospital when an outraged passerby assaulted him for smoking during a blackout.[46] By recoding illumination in terms of necessity, blackout regulations replaced the urban field of luminous legibility with a staccato encounter of distinct classes of licit and illicit light. The blackout led to a new "social geography of lighting," in historian Peter Baldwin's phrase describing the social effects of nineteenth-century losses of artificial lighting.[47]

Inside buildings, especially homes, space was similarly cut up and constricted. Homeowners blacked out rooms, switched off overhead lights, and fitted table lamps with blackout shades to keep light narrowly confined. With windows properly occluded, occupants could use lights in small amounts. Without ambient light, the unitary volume of a room would splinter into discrete pockets of light, each sufficient for one or another particular activity (figure 6.15). Distance disappeared in low or spotty light, and visual objects in turn were perceptually corralled within a relatively small sphere. Correspondingly, authorities recommended specific activities for the narrowed luminous scope: Men could spend their blacked-out evening reading, for example, while women sewed. Children could be amused with board games that likewise needed little light. Home crafts were recommended, such as making blackout lamp shades to control light and to recycle sparse materials. Advocates also believed that by making blackouts the focus of an activity that required creativity and manual dexterity family craft projects could model good citizenship for children (figures 6.16 and 6.17). Quiet introspection was also singled out as a good way to exploit a dim interior, as was conversation. In all such cases, recommended activities for dimly lit rooms involved fine motor skills and small-scale movements, and required little space.

These behavioral recommendations were reinforced by the contemporary scientific literature on scotopic vision. Scientists had already noted that blind subjects who gained or regained eyesight through surgery reported difficulty resolving distances. As one recently sighted man put it, "Objects literally hurled themselves at me."[48] Blumenfeld described similar cases of suddenly sighted patients becoming anxious that visual objects were actually touching their eyes. It was an effect, he wrote, that would be experienced by anyone in a darkened space: "Even to the person with normal sight the dark room seems to extend up to the eyes, whilst its limits tend generally to be 10-40 cm. distant." Citing earlier studies, he noted the consistency of this sensation: to normal eyes a darkened room would appear "absurdly small and near, like the skin of an egg stretched immediately over the eye."[49] Scotopic space, in other words, was characterized by a sense of close vision, echoed in reports of strangers spontaneously coming together around a solitary light in a blackout. While popular sources described the hearthlike attraction of light, it is equally plausible that such ephemeral camaraderie was molded by the constricting circumference of darkness that could press people together.

PATHOLOGIES OF SCOTOPIC SPACE

Blackouts start each day promptly at dusk. Well, you may crow,
Mr. Tojo. You've done a good job of stabbing in the back.
You've darkened our cities.
December 7, 1943[50]

Psychological warfare can be used to enhance panic in enemy country,
to reduce it in friendly country.
Edwin Boring, 1945[51]

The most evident difficulty with diminished lighting outdoors was a sudden rise in the rate of accidents, including everything from stubbed toes to fatal traffic collisions. Daily life was full of mishaps even in the best of circumstances, but lawsuits for blackout-related injuries soared in Britain, and even after the start of World War II there were for some months more fatalities from blackout-related accidents than from active combat.[52] Even in normally lit homes, in 1940, 173,000 people were reportedly killed or disabled by accidents in Britain: "The mistakes made in inky darkness may be tragic," although, with luck, they could be merely "amusing."[53] With forethought, mistakes might also be profitable, noted *Barron's* magazine in a number of articles on blackout investment strategies. Besides betting on greater consumption of liquor, cigarettes (for indoor use), and batteries, investors might consider, for example, buying shares of glass manufacturers, who would enjoy higher sales of replacement windows and automobile headlamps, but they might short sell rubber manufacturers since tires would last longer as drivers stayed home.[54]

While darkened conditions resulted in more accidents, they did not generally encourage deliberate violations or lead to criminal acts in the United States.[55] One 1944 study reviewing criminal statistics from Detroit found that property and personal crimes had not significantly increased during blackouts, even in a city of two million. More common were inadvertent violations of the blackout itself. Violators were reported as being mostly senile or alcoholic, with only eleven deliberate violations, which included two men later accused of spying for the German government and both exhibiting abnormal sexual tendencies. One police chief marveled there were not more violations since the underlying rationale for restrictions was unprecedented and appeared arbitrary: if "in peace time … a community had been told to turn out the lights when a certain whistle blew and to turn them on at another sound, the resistance would have been terrific."[56] The chief articulated something that until the relatively recent development of sensory studies had gone mostly unstudied: societies condition the meanings of sensory stimuli.

But respecting the novel demands of a blackout and inhabiting an unaccustomed space defined by constriction, regulation, and darkness intensified emotional volatility and brought in its wake several pathological conditions, if not actual criminal behavior. The American air raid warden Bill Barbour noted during a blackout that his elderly neighbors left their lights on. He realized that to compensate for their failing eyesight and approaching deafness, they had installed especially bright lights and turned up the volume on the phonograph: "These poor old people, with the dulled senses, still seeking sound, still seeking light!"[57] In their hunger for perceptual stimulation they had missed the siren telling them to turn off the lights and witness the nation's collective rehearsal of their own decaying senses.

If darkness denied sensory stimulation too frequently, it could result in "blackout depression," a condition that British doctors had begun to notice threatened to further aggravate fears of surprise bombardment. The condition was most commonly associated with housewives. One doctor warned that such depression was a greater social ill than the danger of leaving on the lights: "Housewives, with their many anxieties, are being worried and made nervous by repeated calls from wardens who complain even of a glow through dark curtains. … Greater damage may be done to the community in creating neurasthenic states than could result from lighting of this type. … There is ample cause for melancholy, and every stress should be avoided which may intensify depression."[58] In the American setting, not only an unwanted period of darkness could be a potential trigger for depression and anxiety; bright lights could be as well. Efforts to speed factory production involved higher lighting intensities, explained Clarence Ferree and Gertrude Rand, which could prove fatiguing and lead to worker discomfort. Even in homes, they pointed out, a complete blackout during the summer in southern states would be "intolerable."[59]

Men in particular were said to be prone to hysterical night blindness, a sudden failure of an otherwise customary ability to get around in the dark. The syndrome had already been reported during World War I blackouts as a combined product of darkness and nervousness. It was not so much that some men had better night vision than others as that the incidence of night blindness rose during blackouts, in war zones as well as on the home front. Several studies were undertaken to explain the outbreaks. In many cases the defective vision paralleled other physiological or psychological flaws, according to the studies' authors, from degraded nutrition and a decline in levels of vitamin A to a submissive personality, having been reared a "mummy's darling," or harboring homosexual tendencies.[60]

For all clinicians and researchers, the greatest concern was that night blindness was socially infectious; it could spread by panic, which was the most pressing worry associated with American blackouts. When Britons first learned of Germany's rearmament activities, and in the aftermath of Stanley Baldwin's speech about Britain's vulnerability to aerial attack, anxieties spiked in what has since been called "the air panic of 1935."[61] Around 1940, Americans were in a similar position: waiting. As a number of experts attested, such waiting rattled the nerves and led to impatience, tension, poor judgment, and susceptibility to suggestion. Blackouts presented two distinct triggers for panic, first by placing limits on the perception of danger, and second by having even benign objects appear without warning, a condition that was always startling. As one report put it, in the dark, "strange objects have a way of popping suddenly and unexpectedly into 'sight'" and thereby undermining "normal confidence" based on our constant unconscious scanning of surroundings for signs of danger.[62] The eminent experimental psychologist Edwin Boring stated clearly that civilian populations were at risk of panic in these tense situations. He called attention to "mental states" such as anxiety, worry, boredom, and insecurity, whether "imagined or real." All contributed to a vulnerable frame of mind easily tipped into panic by hunger, fatigue, and lack of information, the last of which could result from a sudden decline in sensory input. Worse, as he made clear in a chapter titled "Panic and Mobs," terror was transmissible.[63]

Most vividly, the possibility of panic in the United States had just been demonstrated by Orson Welles's 1938 radio broadcast, "War of the Worlds," which, despite repeated assertions of its fictional premise, led to widespread panic as citizens called one another, worried and confused. Military officials and psychologists such as Boring were still referring to Welles's broadcast into the 1940s. The broadcast had shown several things to analysts: first, that panic could spread almost instantaneously from one person to another, that is, it was a social disease; second, that the modern media of radio and telephone obviated physical proximity as a premise for the spread of panic; and third, that the United States had no accepted civil defense procedures in place in case of an actual attack, alien or otherwise.[64] Thus in many discussions of blackouts, even by those skeptical of an actual aerial attack, blackouts were condoned for contributing to a disciplined citizenry. Lacking prior defense training, civilians needed something to do while they waited for whatever unknowns the war might bring.

To the extent that scotopic space kindled a sense of nearness and encouraged physical proximity, *panic* is an especially apt word. In general terms, it describes a sudden, overwhelming fear, a loss of reason and ruminative faculties, leading to frantic or dangerous actions. But the term also harks back to Pan, the Greek demigod, mischievously fond of suddenly appearing in the dark to surprise travelers and incite them to reckless actions. In this context, Pan could stand as a figure of *Schrecklichkeit*, the German war strategy of inciting fear among civilians through surprise attacks with the ultimate aim of the "obliteration of the individual consciousness." Air raids in particular were understood to be "an attack on the nerves of the civil population" that could dissolve self-control and sever social ties, leaving the populace all the more prone to panic.[65] Studies of the psychology of civilian panic saw blackouts in conflicting terms, as both a psychological self-defense and an incipient form of panic. Even in popular venues, the dialogue of human reason, symbolized as often as not by artificial light, and an irrational nature, betokened by shadow, was often rehearsed. A war-themed poem by Elizabeth Bohm, daughter of the American genre painter Max Bohm, appeared in the *Saturday Evening Post* in 1944, for instance, and concluded: "With darkness come the bird-faced gods and Pan. / A lighted window is the mind of man."[66]

Pan is also a pastoral god of fertility and, under cover of dark, reputed to rise quickly to erotic occasions. Although erotic encounters were hardly pathological, blackouts seem to have brought out this randy side of life as well. As a result of the belief, more than a century in the making, that urban lighting was an inhibiting agent and thus an effective social control, the plunge into darkness was disinhibiting. As the *Washington Post* put it in 1940, "Nobody had a good word to say for the blackout … with the possible exception of the amorous." Scotopic space, in other words, was often sexually charged. A common trope involved a married couple kissing in the confines of a darkened doorway only to discover, or have a passerby discover, that although lawfully wed, they were not wed to each other. Or, to take another example, in 1940 the *Chicago Tribune* ran a short story about

THE SATURDAY EVENING
POST
JUNE 27, 1942
10¢
WHAT TO DO IN A BLACKOUT
property of Willie Gillis, Jr.
Norman Rockwell
THE CASE FOR THE MINORITIES
WENDELL L. WILLKIE
OUR TWO MONTHS ON CORREGIDOR CABOT COVILLE

sabotage and bravery in a London blackout, written by Ethel Lina White, a well-known British crime writer. It is riddled with unexpected, often intimate physical contact and heightened attention to nonvisual sensations. In one passage Christina, the protagonist, climbs aboard a bus during a blackout: "She felt herself borne upwards to the step on a human surge and then pressed forward into a darkened interior. … Helpful hands passed her along the aisle and drew her down on a seat beside a stout woman who smelt strongly of cloves." At other points in the story Christina bumps into shadows or "felt herself gripped by unseen hands."[67] Throughout the story, she becomes an object for strangers' touches even as she bravely pursues an undercover mission. Blackouts inspired several such mystery stories, including a true account of serial murder in 1942, and at least one film. Norman Rockwell's 1942 *Saturday Evening Post* cover illustration—featuring two faces sharply illuminated against a dark background: a woman pressing against a G.I., her hand on his shoulder, her eyes wide to study a booklet he holds titled "What to do in a Blackout," while he glances sideways toward her, a smile forming at the thought of what he would very much like to do in the blackout—struck a perfect balance between a seriousness of civic purpose and an impish impulsiveness that characterized American ambivalence toward blackouts (figures 6.18 and 6.19).[68]

6.18

Ideas of what to do in a blackout differed by the citizen's gender and role in society. Norman Rockwell, *What to Do in a Blackout. Saturday Evening Post*, June 27, 1942, cover. Reproduced by permission of the Norman Rockwell Family Agency. Copyright ©1942 the Norman Rockwell Family Entities.

6.19

The erotic side of blackouts. *What We Found in Last Night's Blackout*, postcard, Ashville Postcard Co., Ashville, North Carolina, ca. 1942. Collection of the author.

Once lights were switched off and the normative counsel of eyesight disabled, the libido erupted into the somatic register in moods that ran from frisky to depraved.

For many, precisely this yielding to those elements of humanity operating outside the deliberative sovereignty of vision underpinned the sense that blackouts dissolved the integrity of civil society. For others, though, the reconfiguration of the senses in scotopic space—its summoning of the full sensorium, as a catalyst for camaraderie, even its ardent ambience—signaled a chance to reinvigorate all those whose instinct and wit had atrophied in the age of advanced sensory aids such as electric light. Where bright lights had formerly turned night into day, the unaccustomed dark sparked new insights and soon bloomed into an all-out, multisensory "taste for the dark," as a 1940 article in the *Washington Post* put it. An ancient "atavistic dread of the dark," still familiar to children and primitive peoples, was suddenly reversed: "Darkness has acquired protective associations something like those the campfire has for the traveler in wild places." Before the war, Americans muddled through the night, blind to its depths. In blackouts, they were "learning the meaning of darkness, that is to say how dark the dark can be. No doubt this overstimulation of our optic nerves has had something to do with what has been called modern temper. ... At any rate, no really good poem about the dawn seems to have been written since the invention of the electric light."[69] In this case, the blackout's violation of a modern sensibility of ubiquitous, even illumination was no throwback to a primitive state but a return to an almost forgotten form of spatial apprehension woven from richly interlacing sensory fibers and the varied visual conditions of the diurnal cycle. Blackouts revealed the underside of electric light. If it was a beacon of mankind's progress, it also entailed the eclipse of nature, a diminishing of the human body's sensory scope and the individual's capacity for wonder.

The best response to blackouts and anxieties arising from anticipating blackouts, according to nearly all commentators on the subject, was to "keep calm and carry on," as the 1939 British Ministry of Information motivational poster suggested. Countless articles appeared early in the war in professional and popular journals advising Americans to follow their customary routines, after, of course, following recommended blackout precautions. Normalcy could keep panic at bay, they said. Lessons for the home front were drawn from psychological studies of panic among soldiers, where poor leadership, indiscipline, anxiety, lack of information, fatigue, and boredom were key contributing factors. Above all, morale needed to be maintained, which meant soldiers, and citizens, should feel secure and engaged with those around them. Good morale, authorities such as Edwin Boring maintained, forestalled panic.[70] For civilian populations in particular, "normal living" was identified as a key area for further research.[71] Based on the English experience, writers stressed the importance of reinforcing social and emotional bonds with others, pointing out that losing a parent is more traumatic to children than enduring an air raid with calm parents at their side.[72]

Motivated by such reasoning, if not simply skepticism about the value of blackout drills in the United States, numerous observers stressed the importance of maintaining routines. Having taken reasonable blackout precautions, Americans could continue with their everyday lives. As one journalist wrote, "It is neither necessary nor desirable to sit in the dark in an atmosphere of nervous tension. The idea is to blackout the enemy not ourselves."[73] To this end, some East Coast towns constructed outdoor blackout curtains along their beaches so that no one would have to forgo an evening stroll along the boardwalk.[74] Families, having sealed their windows against leaking light, learned about activities well suited for an evening together indoors. Journalist Ray Shaw's article "Fun in a Blackout" explained how to make clever "shadowgraphs" on the walls with carefully folded fingers, guaranteeing the activity would "turn a blackout into a family party."[75] Word games were also well suited to blackout conditions.[76] Schools, for their part, kept students busy with assignments to make items such as blackout posters or blackout art they could bring home and use to cover windows to adorn otherwise blacked-out windows.[77] Manufacturers were quick to step in with games to help make blackouts fun. Milton Bradley, which made a number of war-themed games, introduced Blackout in 1939. Players competed, with alternating rolls of the dice, to black out their city window by window (figure 6.20). Retailers capitalized on events by running advertisements about goods for sale in upcoming blackouts, offering tasteful blackout curtains and lamp shades, even special blackout clothing.

LIGHT'S ON

Ironic points of light
Flash out wherever the Just
Exchange their messages:
May I, composed like them
Of Eros and of dust,
Beleaguered by the same
Negation and despair,
Show an affirming flame.

W. H. Auden, 1940[78]

For all the scientific studies, military preparation, civilian training, social organization, home and workplace precautions, envisioned loss of vision, and general hand-wringing, blackouts in the United States were a short-lived phenomenon and, for many, a little silly. Some were simply canceled when citizens objected. The 1940 blackout planned for Kearney, Nebraska, for instance, was called off because "housewives protested it would interfere with [their] electrical refrigerators." As the *Christian Science Monitor* reported, "It could only happen in America."[79] Ward Harrison, a former president of the IES, said that the studies of scotopic vision that dominated lighting research agendas at the start of the war were ill-conceived from the start. Aerial attacks were likely to occur under a full moon, the so-called "bomber's moon," rather than in the dark, so urban patterns would be visible from above despite the most complete blackout. Studying vision in the moonlight would have been a more relevant subject for researchers, he said.[80]

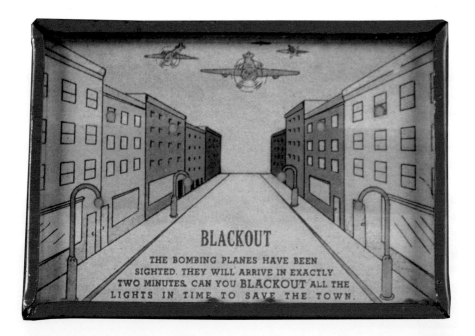

6.20

The enemy bombers are minutes away. Destruction or salvation is in the hands of players of the dexterity game Blackout, who must roll several balls into shallow depressions to obscure light from street lamps and windows. Courtesy of The Strong, Rochester, New York, early 1940s.

As Harrison and many others acknowledged, rehearsing blackouts gave the populace a semblance of security and contributed to a sense of collective mission. It also heightened a sense of national crisis that was, in the end, effective against isolationist sentiment. "Blackout" in turn became a synonym for an anticipated but temporary loss. In 1939, in response to isolationists, who feared a "blackout of prosperity" should the United States engage in war, Franklin Roosevelt proclaimed, "There will be no blackout of peace in the United States," despite the invasion of Poland that had begun just hours before his promise.[81] By early 1941, though, when American participation in the war was looking far more likely, Roosevelt exhorted that the "light of democracy must be kept burning" to avoid the "blackout of barbarism."[82] Against the backdrop of Enlightenment equations of light and human progress, as well as the preceding decades of growing illumination in every area of America life, keeping the lights burning demonstrated courage and determination; phrases such as "the lamps of freedom" thrived.[83] Indeed, a German submarine commander on an espionage mission who ventured into New York Harbor on an un-blacked-out night in 1942, finding his way by means of a guidebook to the 1939 World's Fair, was awed by the city's blazing skyline: lights that burned heedless of the darkness about them could easily be mistaken for lights that burned to defy that darkness.[84] Thus, to long-standing metaphors of light as truth or knowledge or an animating force, a burning electric light came to symbolize collective fortitude.

Crucial to its symbolic weight was that the blackout was an unwanted but finite passage, a condition forced on Americans by wartime threats but that, in due time, would be overcome. "When the lights come on again all over the world," as the 1943 song title had it, spirits would likewise rise. Lights getting switched on became a newsworthy celebration across the nation, once the tide of the war had clearly changed and the mandated precautionary blackouts were finally abandoned with the growing certainty that long-range surprise bombing raids over the American mainland were very unlikely. The representational power of the newly lit light was at its most poignant at the level of the individual, such as the shining eyes of a child gazing in quiet wonder at a Christmas tree, as, for example, in a 1943 General Electric advertisement, which ran with the caption, "The Light no war will ever dim!" (figure 6.21).[85] However much blackouts were a national concern, the struggle with scotopic space was personal, akin to other wartime rationing regimes, a coast-to-coast consensus to forgo bodily comforts in order to achieve a greater good. In turn, in a reciprocal circuit that made the personal political, the challenge to individuals became an allegory of the nation's struggle against fascism: enlisting all the senses, coordinating their actions, and organizing the darkness into a navigable spatial order rehearsed America's own rallying of industrial labor and military service to defeat tyranny.

The patriotic skill in these cases was grounded less in the ability to overcome darkness than in the will to retrain oneself to see in the dark, just as countless men and women retrained to take over essential jobs left open by soldiers fighting overseas. Even more, blackouts engaged the widest range of the citizenry, from children to mothers to the elderly. In this way, mastering the blackout—coordinating disparate messages from sensory outposts in order to form a coherent image of space to get on with the business of winning the war—was seen as an element of character possessed by all those with faith in the legitimacy of their cause. Such individual conviction stood in for national fortitude and often was represented by an inner light, a metaphor that made narrative sense of novel visual conditions. In this way, an imagined perceptual environment was enlisted to serve as an emblem for a national cause.

This point was dramatically expressed on the eve of war by Bette Davis, playing one woman's struggle with oncoming blindness in the film *Dark Victory*. It was adapted from a pre-blackout play that ran on Broadway, but by April 1939, when it was released, and in subsequent adaptations for radio broadcasts throughout the war, the subtext would have been lost on no one as Davis addressed her newly wedded husband: "Nothing can hurt us now. What we have can't be destroyed. That's our victory—our victory over the dark. It is a victory because we're not afraid." Like Davis, all Americans would be able to endure a darkened world as long as their inner values burned on. The light that began to flicker outside was

The Light no war will ever dim!

6.21

The wonder of a child, the spirit of Christmas, and American resolve in wartime, all are equated with a light that never fails. "The light no war will ever dim!" Advertisement, General Electric Co., *Life Magazine* 13, no. 25 (Dec. 21, 1942): 80–81. Reproduced with permission from the Archives and Research Center, Museum of Innovation and Science, Schenectady, New York.

more than matched by unwavering courage that blazed inside. Integrity was depicted as an inner incandescence, which easily eclipsed any technological factor in its making.

In this sense, blackouts in the United States were not the opposite of normal conditions of electric lighting. Rather, they were an episode that demonstrated the extent to which the electric lighting—and the spaces generated by electric lighting—of prior decades had been thoroughly internalized. In cultural terms, then, the American response to wartime blackouts was constructed not as a state of difference from an earlier pre-electric time but as an identification of electric light with a timeless national character. Correspondingly, when the lights did come on again, it was just this penumbra of darkness that had so recently prevailed that, for a time at least, would lend the simple act of switching on a light a momentous weight.

Notes

CHAPTER 1: THE ARCHITECTURE OF ELECTRIC LIGHT

1. Cited in Charles Harrison and Paul Wood, eds., *Art in Theory 1900–1990: An Anthology of Changing Ideas* (Oxford: Blackwell, 1992), 150.

2. Alfred Hay, "Progress in Electrical Engineering," *Electrical Age* 19, no. 5 (Jan. 30, 1897): 70.

3. M. E. Falkus, "The Early Development of the British Gas Industry, 1790–1815," *Economic History Review,* New Series 35, no. 2 (May 1982): 218; William O'Dea, *The Social History of Lighting* (London: Routledge and Kegan, 1958), 1.

4. Falkus, "The Early Development of the British Gas Industry," 219.

5. Maureen Dillon, *Artificial Sunshine: A Social History of Domestic Lighting* (London: National Trust, 2002), 47; Jane Brox, *Brilliant: The Evolution of Artificial Light* (Boston: Houghton Mifflin, 2010), 77–78.

6. Brox, *Brilliant,* 78–81; Roger Ekirch, *At Day's Close: Night in Times Past* (New York: Norton, 2006), 163.

7. Much of the arc lamp's success was due to the recent availability of a steady source of power, the electric dynamo. In 1832 the English scientist Michael Faraday formulated the principles of electromagnetic induction, according to which an electric circuit may interact with a magnetic field to produce electricity. Over the next thirty years inventors investigated ways to produce a current that would be strong and steady enough to use for industrial purposes, such as powering electric arc furnaces to refine metals. Zénobe Gramme, a Belgian electrical engineer who had invented the first commercially viable dynamo in 1871, collaborated with Yablochkov on the Paris lights. In the early 1880s the American engineer and inventor Charles Brush installed thousands of arc lamps of his own design, powered by his own patented dynamos, in dozens of cities across the United States. See Matthew Luckiesh, "A Half-Century of Artificial Lighting," *Industrial and Engineering Chemistry* 18, no. 9 (September 1926): 920.

8. In 1929 the United States Postal Service celebrated "Electric Light's Golden Jubilee" with a postage stamp. There are many titles that tell the story of electricity and electric lighting. For several recent ones, see Alice Barnaby, *Light Touches: Cultural Practices of Illumination* (London: Routledge, 2017); Tim Edensor, *From Light to Dark: Daylight, Illumination, and Gloom* (Minneapolis: University of Minnesota Press, 2017); and Ernest Freeberg, *The Age of Edison: Electric Light and the Invention of Modern America* (New York: Penguin Press, 2013).

9. Margaret Maile Petty, "Cultures of Light: Electric Light in the United States, 1890s–1950s" (Ph.D. diss., Victoria University of Wellington, 2016), 212–221.

10. Samuel G. Hibben, "The Society's First Year," *Illumination Engineering* 51, no. 1 (June 1956): 145–150. See the discussion in David L. DiLaura, *A History of Light and Lighting: In Celebration of the Centenary of the Illuminating Engineering Society of North America* (New York: Illuminating Engineering Society of North America, 2006), 7–9.

11. Noted in Edward Graham Daves Rossell, "Compelling Vision: From Electric Light to Illuminating Engineering 1880–1940" (Ph.D. diss., University of California, 1998), xix. See also

Bassett Jones, Jr., "The Relation of Architectural Principles to Illuminating Engineering Practice," *TIES (Transactions of the Illuminating Engineering Society)* 3, no. 1 (January 1908): 9.

12. DiLaura, *A History of Light and Lighting*, 19; and David DiLaura "It Has Been Proposed to Form a Society," *Lighting Design and Application* 36, no. 1 (2006): 45. See also Sean F. Johnston, *A History of Light and Color Measurement: Science in the Shadows* (London: IOP Publishing, 2001), 84.

13. E. L. Elliott, "Engineer or Architect," editorial, *Illuminating Engineer* 2, no. 5 (July 1907): 385.

14. E. L. Elliott, "Electric Light as Related to Architecture," editorial, *Illuminating Engineer* 2, no. 7 (September 1907): 525.

15. For more on the formation of the lighting design profession, see Petty, "Cultures of Light," and Rossell, "Compelling Vision."

16. For more on the tension between architects and illuminating engineers, see Sandy Isenstadt, "Architects and Artificial Light," *Histoire de l'énergie / History of Energy*, ed. Comité d'histoire de l'électricité et de l'énergie (Alain Beltran, Léonard Laborie, Pierre Lanthier, and Stéphanie Le Gallic) (Frankfurt am Main: Peter Lang, 2016), 81–102.

17. Cited in Harrison and Wood, eds., *Art in Theory 1900–1990,* 169.

18. Vladimir Mayakovsky, "Broadway," in *Selected Poems,* trans. J. H. McGavran (Evanston: Northwestern University Press, 2013), 112; Vladimir Mayakovsky, "My Discovery of America (1925–26)," in *America through Russian Eyes, 1874–1926,* ed. Olga Peters Hasty and Susanne Fusso (New Haven: Yale University Press, 1988), 171.

19. Vladimir Mayakovsky, "Conversation with a Taxman about Poetry," in *Selected Poems,* 135.

20. Winifred Lewellin James, *A Woman in the Wilderness* (New York: George H. Doran, 1916), 13–17.

21. Franklin Leonard Pope, *Evolution of the Electric Incandescent Lamp* (Elizabeth, NJ: Cook and Hall, 1889), 39.

22. James Clerk Maxwell, *A Treatise on Electricity and Magnetism*, vol. 2, 3rd ed. (New York: Dover, 1954), 493. See the discussion in Mary B. Hesse, *Forces and Fields: The Concept of Action at a Distance in the History of Physics* (London: Thomas Nelson and Sons, 1961), 195–198, and throughout on action at a distance generally. On period responses, see "Action at a Distance," *Cassier's* 42, no. 2 (August 1912): 164–166, and "Action at a Distance," *Scientific American* 77 (Jan. 17, 1914): 39.

23. Kern argued that telegraphy was the start of a sense of universal time antedating the establishment of World Standard Time in 1884. Stephen Kern, *The Culture of Time and Space 1880–1918* (Cambridge, MA: Harvard University Press, 1983), 314. See also Iwan Rhys Morus: "The Electric Ariel: Telegraphy and Commercial Culture in Early Victorian England," *Victorian Studies* 39, no. 3 (Spring 1996): 339–378; and "'The Nervous System of Britain': Space, Time and the Electric Telegraph in the Victorian Age," *British Journal for the History of Science* 33, no. 4 (December 2000): 455–475.

24. Robert Louis Stevenson, "A Plea for Gas Lamps," in *Virginibus Puerisque* (New York: Charles Scribner's Sons, 1887), 277.

25. *Fundamentals of Illumination Design* (Cleveland, OH: Engineering Department, National Lamp Works of General Electric Co., Dec. 20, 1917), 35–36. See also "Illumination Terms," *Lighting Data Bulletin* LD 155 (Harrison, NJ: Edison Lamp Works of General Electric Co., n.d.), 25.

26. Charles P. Steinmetz, "Light and Illumination," *Scientific American* 63, Suppl. 1630 (Mar. 30, 1907): 26114. Read before a meeting of the IES in December 1906 and published widely afterward.

27. P. W. Cobb and F. K. Moss, "Glare and the Four Fundamental Factors in Vision," *TIES (Transactions of the Illuminating Engineering Society)* 23 (September 1928): 1104.

28. Wolfgang Schivelbusch describes the arc light as "a small sun" and "an artificial sun": *Disenchanted Night: The Industrialization of Light in the Nineteenth Century*, trans. Angela Davies (Berkeley: University of California Press, 1995), 118–120.

29. Émile Zola, *The Ladies' Paradise: A Realistic Novel,* trans. Ernest Alfred Vizetelly (London: Vizetelly and Co., 1886), 377.

30. Gösta Bergman, *Lighting in the Theater* (Totowa, NJ: Rowman and Littlefield, 1977), 287.

31. As Edgar Allan Poe lamented, glare was already "a leading error in the philosophy of American household decoration" under conditions of gas lighting. Edgar Allan Poe, "The Philosophy of Furniture," *Burton's Gentleman's Magazine* 6, no. 5 (May 1840): 243–245, at 244.

32. "The Development of Electric Lighting," *Quarterly Review* 152 (October 1881): 459.

33. "Fixture, Lighting, Auditorium," in *EMF Electrical Year Book: An Encyclopedia of Current Information about Each Branch of the Electrical Industry*, ed. Frank H. Bernhard (Chicago: Electrical Trade Publishing Co., 1921), 279.

34. Albert Scheible, "Illumination vs. Glare," *Electrical Engineer* 20, no. 397 (December 11, 1895): 565–566.

35. Michel Foucault, *Madness and Civilization: A History of Insanity in the Age of Reason*, trans. Richard Howard (New York: Routledge, 2001), 101–105. [Emphasis in original.]

36. Marshall McLuhan, *Understanding Media* (New York: McGraw-Hill, 1964), 3.

CHAPTER 2: AT THE FLIP OF A SWITCH

1. Seamus Heaney, *Electric Light: Poems* (New York: Farrar, Straus, 2002), 96.

2. Kenneth Serbin, "Church-State Reciprocity in Contemporary Brazil: The Convening of the International Eucharistic Congress of 1955 in Rio de Janeiro," *Hispanic American Historical Review* 76, no. 4 (November 1996): 726–727.

3. Elizabeth Bishop, "On the Railroad Named Delight," *New York Times*, Mar. 7, 1965, SM30.

4. It was not for nothing that in 1909 General Electric named its new line of tungsten filament bulbs Mazda, after the Zoroastrian god affiliated with light.

5. "America's Electrical Week and Plans for the Holidays," *Electrical Review and Western Electrician* 69, no. 19 (Nov. 4, 1916): 797–803, at 801; and George Buchanan, "Aladdin's Lamp Outdone," *Bell Telephone News* 11, no. 1 (August 1921): 35.

6. Edward De Segundo, "Electric Light and Gas," *Electrical Review* 32, no. 792 (Jan. 27, 1893): 89.

7. Bennet Woodcroft and Translator, *The Pneumatics of Hero of Alexandria* (London: Taylor Walton and Maberly, 1851), 96–100. I thank Ruth Bielfeldt for this reference.

8. "Lane Fox v. The Kensington and Knightsbridge Electric Lighting Company," *Reports of Patent, Design and Trade Mark Cases* 9, no. 23 (June 22, 1892): 221–249, at 235.

9. For an elaboration, see G. Claude, "Electrical Phenomena Illustrated by Applied Hydraulics," *Scientific American Supplement* 38, no. 979 (Oct. 6, 1894): 15652–15654.

10. The German word for switch, *Schalter*, is etymologically related both to a rod or bar used to push something away and to an opening that can be closed by a shutter or sliding panel. The French term, *robinet*, also used in plumbing applications and translated as tap or faucet, is from Latin. See Iwan Rhys Morus, "The Electric Ariel: Telegraphy and Commercial Culture in Early

Victorian England," *Victorian Studies* 39, no. 3 (1996): 339–378; and Iwan Rhys Morus, "'The Nervous System of Britain': Space, Time and the Electric Telegraph in the Victorian Age," *British Journal for the History of Science* 33, no. 4 (Dec. 1, 2000): 455–475.

11. Other means of igniting sources are described in Maureen Dillon, *Artificial Sunshine: A Social History of Domestic Lighting* (London: National Trust, 2002), 33; and Jane Brox, *Brilliant: The Evolution of Artificial Light* (Boston: Houghton Mifflin, 2010), 63–65.

12. Further details of gas lamp operation are described in Chris Otter, *The Victorian Eye: A Political History of Light and Vision in Britain, 1800–1910* (Chicago: University of Chicago Press, 2008), 234–235.

13. "A Dazzling Exhibition," *New York Times*, Sept. 3, 1884, 1.

14. Theater lighting, where the technology and theory of switching and dimming was most advanced, is beyond the scope of this chapter. Sources include Theodore Fuchs, *Stage Lighting* (New York: Blom, 1929); Louis Hartmann, *Theater Lighting: A Manual of the Stage Switchboard* (New York: D. Appleton, 1930); Stanley McCandless, *A Method of Lighting the Stage* (New York: Theater Arts, 1932); L. G. Applebee, "The Evolution of State Lighting," *Journal of the Royal Society of Arts* 94, no. 4723 (Aug. 2, 1946): 550–563; Barnard Hewitt, ed., *The Renaissance Stage: Documents of Serlio, Sabbattini and Furttenbach* (Coral Gables, FL: University of Miami Press, 1958); Donald L. Murray, *The Rise of the American Professional Lighting Designer to 1960* (Ann Arbor: University of Michigan Press, 1970); and Jody Briggs, *Encyclopedia of Stage Lighting* (Jefferson, NC: McFarland, 2003).

15. Thomas Edison, Electric Lamp and Holder for the Same, US265,311 A, filed Feb. 5, 1880, and issued Oct. 3, 1882 (http://www.google.com/patents/US265311).

16. Thomas A. Edison, System of Electric Lighting, US251551 A, filed Aug. 30, 1881, and issued Dec. 27, 1881 (http://www.google.com/patents/US251551). See also T. P. Hughes, "Thomas Alva Edison and the Rise of Electricity," in *Technology in America*, ed. Carroll Pursell, Jr. (Cambridge, MA: MIT Press, 1981), 123–125.

17. "Gas Lighting by Electricity," *Engineering* 44 (July 8, 1887): 53.

18. W. J. Keenan and James Riley, *The Transmitted Word* (Boston: Dorchester Press, 1897), vii.

19. See Carolyn Marvin, *When Old Technologies Were New* (New York: Oxford University Press, 1990), 119–122. Regarding the dangers of wires, see, for example, "Sudden Death in the Air: Dangers of Electric Arc Lights and Storage Batteries" *New York Times*, Nov. 4, 1883, 6.

20. John Trowbridge, *What Is Electricity?*, International Scientific Series 75 (New York: Appleton and Co., 1896), 308–309. See also the discussion titled "The Uncertain Identity of Electricity" in Graeme Gooday, *Domesticating Electricity: Technology, Uncertainty and Gender, 1880–1914* (London: Pickering and Chatto, 2008).

21. James Thurber, "My Life and Hard Times: The Car We Had to Push," *New Yorker*, July 15, 1933.

22. A. E. Kennelly, "Electricity in the Household," *Scribner's Magazine* 7 (Jan. 1, 1890): 107.

23. Robert Hammond, *The Electric Light in Our Homes* (London: Frederick Warne and Co., 1884), 82, 85–86. Another example: "Every room will have a switch placed beside the doorway, so that the lamps can be most conveniently controlled by a person entering or leaving the room. In such a house the most timid child, of but a few years, will go anywhere with safety and without fear, lighting and extinguishing the lamps as desired." *Electricity in a Modern Residence* (New York: H. Ward Leonard and Co., 1892), 5.

24. Sigmund Freud, "Beyond the Pleasure Principle" (1920), in *The Standard Edition of the Complete Psychological Works of Sigmund Freud*, vol. 18, trans. James Strachey (London: Hogarth Press, 1955), 14–17.

25. Later enhancements such as motion detectors or timers are simply automatic means of ensuring that the user's desire for light is satisfied, buttons pressed in advance, so to speak.

26. Kennelly, "Electricity in the Household," 110.

27. "Miscellaneous City News," *New York Times*, Sept. 5, 1882, 8.

28. "Touching the Button," *Independent*, May 11, 1893, 10.

29. "Opening the World's Fair," *Electrical Engineer* 15, no. 261 (May 3, 1893): 439. The Corliss engine was the most efficient stationary steam engine of the nineteenth century. Typically, as in Chicago, it was belted to drive dynamos and produce electricity.

30. "The President Presses the Button," *New York Times*, May 2, 1893, 1–2.

31. "Opened by the President," *New York Times*, May 2, 1893, 1.

32. "Touching the Button," 10.

33. Ibid.

34. Ibid.

35. Charles Edward Bolton, *A Model Village of Homes and Other Papers* (Boston: L. C. Page and Co., 1901), 187–188.

36. See Colin Ford and Karl Steinorth, eds., *You Press the Button, We Do the Rest: The Birth of Snapshot Photography* (London: Nishen, 1988). The camera was marketed especially to women, who proved to be avid consumers of it, although a number of articles suggested they underestimated the discipline needed for quality work and were prone to mistakes. "Women Who Press the Button," *New York Times*, Oct. 1, 1893, 18.

37. "The Age of Buttons," *Chicago Daily Tribune*, Jan. 8, 1900, 6.

38. The Portland speech, September 21, 1932. Roosevelt employed the switch metaphor here to call for greater financial transparency in the electrical industry. Craig Roach, *Simply Electrifying: The Technology That Transformed the World, from Benjamin Franklin to Elon Musk* (Dallas, TX: BenBella Books, 2017), 135–136.

39. "At President's Touch: Spark from the White House Lights Christmas Tree," *Washington Post*, Nov. 24, 1903, 2; "Gates of St. Louis Big Exposition Thrown Open to Nations of Earth," *Atlanta Constitution*, May 1, 1904, A7; "Coolidge Lights Up Tree to Inaugurate Nation's Christmas," *New York Times*, Dec. 25, 1925, 1; "City's Yule Tree to Gleam Tonight at Coolidge's Touch," *Washington Post*, Dec. 24, 1927, 1.

40. "President to Inaugurate Christmas Cheer Season," *Washington Post*, Dec. 16, 1929, 10; "Hoover Speaks on Radio Today," *Los Angeles Times*, Dec. 24, 1930, A11.

41. "Throngs Attend Ceremonies at Christmas Tree," *Chicago Daily Tribune*, Dec. 25, 1916, 3; "Lights Blaze from Big Tree," *Boston Daily Globe*, Dec. 24, 1922, 1, 14; "Mayor McGrath Switches on Christmas Lights at Quincy," *Daily Boston Globe*, Dec. 13, 1930, 12; "Fairies to Light Christmas Trees," *Los Angeles Times*, Dec. 17, 1937, A2; "City Lights Trees in Every Borough: As His Daughter Pulls Switch," *New York Times*, Dec. 21, 1938, 26; "City's Yule Fetes to Begin Tuesday," *New York Times*, Dec. 18, 1938, 24; "Christmas Trees in Parks Lighted," *New York Times*, Dec. 21, 1939, 21.

42. "Miss Hightower 'Pressed the Button,'" *Electrical Engineer* 23, no. 464 (Mar. 24, 1897): 314.

43. Arnaldo Cortesi, "New Power Plant Opened by Pontiff," *New York Times*, Feb. 7, 1931, 7.

44. "Inauguration of Electric Service Is Occasion for Big Celebration on the Part of Many Municipalities," *National Electric Light Association Bulletin* 9, no. 8 (August 1922): 483.

45. "Barnesville: Roosevelt Forgets a Chore," *New York Times*, Aug. 13, 1938, 3; "Georgians Await Roosevelt Word," *New York Times*, Aug. 7, 1938, 2; Ralph McGill, "One Word More," *Atlanta Constitution*, June 23, 1939, 10.

46. "President Opens Boulder Dam by Pressing Button," *Christian Science Monitor*, Sept. 11, 1936, 1, 11; "The Text of President Roosevelt's Address at World Power Conference," *New York Times*, Sept. 12, 1936, 3. See also "Third Power, Second Dams," *Time* 28, no. 12 (Sept. 21, 1936): 76–78.

47. On radio's creation of an American public, see Jason Loviglio, *Radio's Intimate Public: Network Broadcasting and Mass-Mediated Democracy* (Minneapolis: University of Minnesota Press, 2005); and Susan Douglas, *Listening In: Radio and the American Imagination* (Minneapolis: University of Minnesota Press, 2004). I thank David Suisman for these references.

48. Francis Trevelyan Miller, *America the Land We Love* (New York: Search-Light Book Corp., 1915), 217–218.

49. The light switch joins other small gestures of modernity such as striking a match, lifting a telephone receiver, and pressing a camera button in that, as Walter Benjamin put it, "one abrupt movement of the hand triggers a process of many steps." "On Some Motifs in Baudelaire," in *Illuminations: Essays and Reflections*, ed. Hannah Arendt, trans. Harry Zohn (New York: Schocken, 1968), 174–175.

50. "Splendid Night Display," *Los Angeles Times*, Aug. 12, 1902, 1.

51. Ibid.

52. "Opened by the President," *New York Times*, May 2, 1893, 1.

53. Albert Borgmann, *Technology and the Character of Contemporary Life* (Chicago: University of Chicago Press, 1984), 41. Device paradigms are always culturally perceived in relation to alternatives. Thus gas lamps were felt to deliver their commodity without mediation. See, for example, Friedrich Accum, *Practical Treatise on Gas-Light* (London: G. Hayden, 1815), 6.

54. Borgmann, *Technology and the Character of Contemporary Life*, 125. The element that measures the flow of benefits, the electric meter, is hidden from the consumer. Borgmann suggested that the device paradigm resulted from the separation of producers from consumers and the formation of consumer society in general.

55. Switches might even more broadly be understood as handles by which we gain leverage on the material reality of technology, between ourselves and our technologically mediated habitats.

56. A. E. Kennelly, "Electricity in the Household," *Scribner's Magazine* 7, no. 1 (January 1890): 107.

57. "Touching the Button," *Independent* 45, no. 2319 (May 11, 1893): 10.

58. C. Mackechnie Jarvis, "The Distribution and Utilization of Electricity," in *A History of Technology*, ed. Charles Singer et al. (Oxford: Clarendon Press, 1958), 5:230–231.

59. "The Brownie in the Kitchen," *Edison Monthly* 12, no. 2 (February 1920): 42.

60. "The Age of Buttons," *Chicago Daily Tribune*, Jan. 8, 1900, 6.

61. "The House of 100 Years from Now," *St. Louis Post-Dispatch*, Oct. 20, 1907, B2.

62. *The Electric House*, directed by Edward F. Cline and Buster Keaton (First National Pictures, 1922). On the formation of technological slapstick more generally, see Rob King, "'Uproarious Inventions': The Keystone Film Company, Modernity, and the Art of the Motor," *Film History* 19 (2007): 271–291.

63. Arthur Stringer, *The Wire Tappers* (Boston: Little, Brown, 1906), 187, is discussed along with other examples in Robert MacDougall, "The Wire Devils: Pulp Thrillers, the Telephone, and

Action at a Distance in the Wiring of a Nation," *American Quarterly* 58, no. 3 (September 2006): 715–741. For other examples of dramatic switches, see Jack London, "To Kill a Man" and "When the World Was Young," in *The Night Born* (New York: Grosset and Dunlap, 1913).

64. Alice Carroll, "Render Service after the Sale," *Electragist* 22, no. 10 (August 1923): 17.

65. Alice L. James, *Catering for Two: Comfort and Economy for Small Households* (New York: Putnam's Sons, 1898), 317.

66. "Presto! Dinner Is Ready," *New York Times*, April 10, 1894, 9.

67. Ibid.

68. For more on so-called labor-saving devices, see the classic book, Ruth Schwartz Cowan, *More Work for Mother: The Ironies of Household Technology from the Open Hearth to the Microwave* (New York: Basic Books, 1983).

69. "An Automatic Housekeeper," *Washington Post*, Mar. 1, 1908, M3.

70. "In the Night Watches: An Idyl of the Fin de Siecle," *Atlanta Constitution*, July 27, 1895, 8.

71. Discussion following Mathias E. Turner, "Report on Electric Heating Devices" delivered before the Ohio Electric Light Association. Douglas Brown, Official Reporter, *Proceedings of Ohio Electric Light Association* 13 (Cincinnati, 1907), 49.

72. See, to name only one example, "Housewifery Made Easy by Electrical Devices," *Washington Post*, Feb. 25, 1906, SM5.

73. "Electricity Does Home Work," *Arizona Republican*, May 2, 1911, A1.

74. Ironically, women were often responsible for assembling electrical equipment such as switches. Women's smaller fingers were said to be better suited for the fine work involved in making switches and sockets and, despite their gender, they were in industry reports fully capable of operating power machinery. More likely they were employed because they accepted lower wages than men and Progressive Era safety reforms along with automation had made factories safer for all workers. Charles M. Ripley, *Romance of a Great Factory* (Schenectady, NY: Gazette Press, 1919), 147–149. Ripley was both an engineer and a leading publicist for General Electric and its efforts to portray the firm as a family. See Julia Kirk Blackwelder, *Electric City: General Electric in Schenectady* (College Station: Texas A&M University Press, 2014), 34–35.

75. "The Pushbutton Habit," *Electric Journal* 14, no. 2 (February 1917).

76. Carolyn M. Goldstein, *Creating Consumers: Home Economists in Twentieth-Century America* (Chapel Hill: University of North Carolina Press, 2012), 212–221.

77. Mrs. C. B. Tate, Jr., "The Field of Women in Public Relations," *N.E.L.A. Bulletin* 9, no. 4 (April 1922): 249–250.

78. Frank Lloyd Wright, *The Disappearing City* (New York: William Farquhar Payson, 1932), 3.

79. Martin Heidegger, "The Question Concerning Technology," in *The Question Concerning Technology and Other Essays*, trans. William Lovitt (New York: Garland Publishing, 1977), 3–35.

80. Henry Adams, *The Education of Henry Adams* (Boston: Houghton Mifflin, 1918), 380–381.

81. Karl Heinz Bohrer, "Instants of Diminishing Representation: The Problem of Temporal Modalities," in *The Moment: Time and Rupture in Modern Thought*, ed. Heidrun Friese (Liverpool: Liverpool University Press, 2001), 115.

82. Introduced in Fredric Jameson, *Postmodernism; Or, The Cultural Logic of Late Capitalism* (Durham, NC: Duke University Press, 1991), 35–37, and elaborated in David E. Nye, *American Technological Sublime* (Cambridge, MA: MIT Press, 1994).

83. Sigmund Freud, "The Uncanny" (1919), in *The Standard Edition of the Complete Psychological Works of Sigmund Freud*, vol. 17, trans. James Strachey (London: Hogarth Press, 1955), 237.

CHAPTER 3: DRIVING THROUGH THE AMERICAN NIGHT

An earlier version of this chapter appeared as "Auto-Specularity. Driving through the American Night," *Modernism/Modernity* 18, no. 2 (Spring, 2011): 213–231.

1. James Jerome Gibson and Laurence E. Crooks, "A Theoretical Field-Analysis of Automobile-Driving," *American Journal of Psychology* 51, no. 3 (July 1938): 453.

2. Edith Louise Coues O'Shaughnessy, *Diary of Edith Louise O'Shaughnessy* (New York: Harper and Brothers, 1918), 130.

3. "Letter from Sarah Diodati Gardiner, July 28, 1908," in *Pages in Azure and Gold: The Letters of Miss Gardiner and Miss Quincy* (Privately published, 1915), 247.

4. A useful overview of automobile lighting is Peter W. Card, *Early Vehicle Lighting* (Aylesbury: Shire, 1987).

5. Benjamin Avery, "Summering in the Sierra, No. II," *Overland Monthly and Out West Magazine* 12, no. 2 (February 1874): 175–183, at 176.

6. Henry van Dyke, "Silverhorns," *Scribner's* 41, no. 4 (April 1907): 429–437.

7. "Railroad Transportation," *Scientific American* 87, no. 24 (Dec. 13, 1902): 398.

8. Ibid.

9. On electric cars more generally, see James Flink, *The Automobile Age* (Cambridge, MA: MIT Press, 1988), 8–10.

10. Table Q in *Historical Statistics of the United States* (Washington, DC: U.S. Census Bureau, 1975), 148–162.

11. "Auto Club Denounces Many Motor Evils," *New York Times*, Dec. 13, 1906, 7.

12. "The New Juggernaut," *Washington Post*, Jan. 3, 1900, 2.

13. Harry X. Cline, "Electric Lights Make Night Driving a Pleasure," *Journal of the American Medical Association* 58, no. 14 (Apr. 6, 1912): 1072.

14. "Familiar Objects Are Often Unseen," *Boston Daily Globe*, Mar. 6, 1910, 60.

15. "Automobile Notes," *Manchester Guardian*, Nov. 20, 1908, 4.

16. Harold Whiting Slauson, "New Aids for the Motorist," *Outing*, March 1911, 759. My thanks to Basie Gitlin for this reference.

17. Harold Whiting Slauson, "Ways That Are Dark," *Atlanta Constitution*, Oct. 1, 1911, D7.

18. See "Proximate Cause and Contributory Negligence of Traveler," in *Abbott's Digest of All New York Reports*, ed. George F. Longsdorf (New York: Baker, Voorhis and Co. and Lawyers Co-Op Publishing Co., 1915), 18:1257–1261.

19. E. A. Greathed, "Some Aspects of the Automobile," *Monthly Review* 19, no. 56 (May 1905): 59–60.

20. "Night Driving," *Los Angeles Times*, Apr. 3, 1910, VII3.

21. "Night Auto-Driving," *Hartford Courant*, May 15, 1906, 4.

22. Jonathan Crary, *Suspensions of Perception: Attention, Spectacle, and Modern Culture* (Cambridge, MA: MIT Press, 2001), 10.

23. "Banish the Dazzling Headlight," letter to the editor, *Chicago Daily Tribune*, June 28, 1913, 6. A number of studies considered whether the urban or the rural nighttime visual field was more complex and therefore more demanding on drivers. See "Report of the 1916–17 Committee on Automobile Headlamps," *Transactions of the Illuminating Engineering Society* 13, no. 5 (July 20, 1918): 267.

24. Slauson, "Ways That Are Dark."

25. "Night Driving."

26. "Careful, Mr. 'Joy Rider,' the 'Night Riders' Will Get You If You Don't Watch Out," *Chicago Daily Tribune*, July 25, 1909, I1.

27. Cited in Michael L. Bromley, *William Howard Taft and the First Motoring Presidency, 1909–1913* (Jefferson, NC: McFarland and Co., 2003), 95. See also Michael L. Berger, *The Devil Wagon in God's Country: The Automobile and Social Change in Rural America, 1893–1929* (Hamden, CT: Archon Books, 1979), 68.

28. Henri Lefebvre, *The Production of Space*, trans. Donald Nicholson-Smith (Oxford: Blackwell, 1991), 25.

29. See Stephen Kern, *The Culture of Time and Space: 1880–1918* (Cambridge, MA: Harvard University Press, 1983); and David Harvey, "The Experience of Space and Time," in *The Condition of Postmodernity* (Cambridge, MA: Blackwell, 1990), 201–326.

30. "The Weakest Link," *Street Railway Journal* 26, no. 9 (Aug. 26, 1905): 297. Railroad engineers faced an additional matter of having to account for a train's length since the first and last cars would clear a particular point at different times. See "The ABC of Railroad Signaling," *Engineering Digest* 6 (December 1909): 508–516, at 510–11.

31. "Non-Glaring Headlight Discussion," *Society of Automobile Engineers Transactions* 9, no. 2 (1914): 284–286.

32. Clayton H Sharp, "Safety Devices Necessary," *Motordom* 13, no. 11 (May 1918): 13.

33. Ibid., 13.

34. Adrian Delsman et al., Appellants, v. Albert Bertotti, Respondent, No. 27551, S. Ct. Wash., Dept. 2, 200 Wash. 380; 93 P.2d 371; 1939 Wash. LEXIS 448, August 31, 1939; Sallie B. Temple and Others, Administrators v. John A. Moses, No. 2178 S. Ct. Va., 175 Va. 320; 8 S.E.2d 262; 1940 Va. LEXIS 175, April 8, 1940.

35. Arthur Leff, "The Leff Dictionary of Law: A Fragment" (1855), *Yale Law Journal* 94 (July 1985): 2086.

36. See, for example, Roy Welday, *Your Automobile and You* (New York: Henry Holt, 1938), 74.

37. See, for example, "Show Aldermen Some Quick Stops," *Automobile Topics* 26, no. 2 (May 25, 1912): 98.

38. "Safety Code for Automobile Brakes," *Science*, New Series, 57, no. 1478 (Apr. 27, 1923): 493.

39. Gibson and Crooks, "A Theoretical Field-Analysis of Automobile Driving," 453–471; 457–458.

40. Evan Edwards and H. H. Magdsick, *Light Projection: Its Applications* (Cleveland, OH: National Lamp Works of General Electric Co., Engineering Dept., Aug. 25, 1917): 11.

41. Heldt, "The Problem of Eliminating Headlight Glare," 400.

42. Ibid.

43. J. B. Replogle, "The Determination of Headlight Glare," *Automobile* 35, Nov. 23, 1916, 890–892, at 890.

44. "Headlights a Danger," *New York Times*, Apr. 19, 1914, XX9. Slowing down would obviate brighter headlights altogether, a point voiced by many. See, for example, "The New Night Terror," letter, *British Medical Journal*, Aug. 26, 1911, 464.

45. "Headlights a Danger."

46. Edwards and Magdsick, *Light Projection*, 11.

47. James Kerr, "Glare and the Student's Life," *Mind & Body* 26, no. 282 (November 1919): 241–248, at 245.

48. H. Massac Buist, "Motor Notes for Medical Men," *British Medical Journal* 1, no. 3094 (Apr. 17, 1920): 543–545, at 543.

49. "'Tis a Fast Age," *Los Angeles Times*, Sept. 27, 1903, A4.

50. "The Automobile As an Animal," *New York Times*, Apr. 14, 1904, 8. Train headlamps had been similarly described: "a blind monster, but for the huge cyclop eye fixed in his forefront," in "Hamblen's Locomotive Headlights, &c," *Chicago Press and Tribune*, June 1, 1859. Thanks to Basie Gitlin for this reference.

51. Claude Perrin Berry, *The Law of Automobiles*, 3rd ed. (Chicago: Callaghan and Co., 1921), 21.

52. "Speed Maniacs and Road Hogs," *Los Angeles Times*, Dec. 27, 1908, VI3, is one of many examples.

53. See, for example, "Dangerous Headlights," *Washington Post*, June 19, 1908, 6; or "Headlight Causes a Crash'" *Chicago Daily Tribune*, July 31, 1913, 3. Sometimes the animals fought back. In one instance a bull moose "was enraged by the strong glare of the lights" so much that it "made a savage rush for the car." The car's occupants, returning from Waterville, Maine, were unharmed. See "Angry Moose Charges Auto," *Chicago Defender*, Dec. 26, 1914, 3.

54. "Cities Newest Terror the Night Riding Auto," *Chicago Daily Tribune*, May 7, 1911, G6.

55. Ibid.

56. Ibid.

57. Ibid.

58. "Two Typical 'Joy Rides,'" *New York Tribune*, Nov. 11, 1909, 6.

59. "Automobile Notes," *Manchester Guardian*, Nov. 20, 1908, 4.

60. John R. Eustis, "Haste Makes Waste," *Independent* 94, no. 3422 (May 4, 1918): 216.

61. "The Occultation of Headlights," *Automotor Journal (London)* 9, no. 11 (Mar. 3, 1906): 251.

62. "Lamps and Ladies: Dangerous Obstacles," *Motor* 4, no. 95 (Dec. 2, 1903): 442.

63. Jonathon Green, "headlight," in *Green's Dictionary of Slang* (Chambers Harrap Publishers, 2010) (http://proxy.library.upenn.edu:2514/view/10.1093/acref/9780199829941.001.0001/acref-9780199829941-e-22477).

64. *They Drive by Night*, directed by Raoul Walsh (Warner Bros., 1940).

65. "Use All the Vision," *Los Angeles Times*, Mar. 27, 1932, F4.

66. "Psychologist Informs Physicians Auto Crashes Are Caused by Limited Sight Field," *New York Times*, Apr. 18, 1936, 17.

67. James Colbert, "Woman Driver Tested," *New York Times*, Mar. 8, 1936, XX12.

68. "Women More Susceptible to Glare at Night," *Science News-Letter* 40, no. 14 (Oct. 4, 1941): 214–215.

69. Edith Eugenie Johnson, *Leaves from a Doctor's Diary* (Palo Alto, CA: Pacific Books, 1954), 35.

70. "Banish the Dazzling Headlight," letter to the editor, *Chicago Daily Tribune*, June 28, 1913, 6.

71. L.B. Marks, "Inaugural Address," *TIES (Transactions of the Illuminating Engineering Society)* 1, no. 1 (February 1906): 4–6.

72. Charles G. MacLeod and Ch. M. Edin, "Electric Light Asthenopia," *The Australasian Medical Gazette*, 12 (June 1893): 196; Thomas R. Pooley, "Asthenopia not Dependent upon Errors of Refraction or Insufficiency of the Ocular Muscles," *Transaction of the New York Academy of Medicine*, 10 (1894): 613; S. Busby Allen, Asthenopia Due to Glare, *Manhattan Eye and Ear Hospital Reports*, 3 (Jan. 1896): 50–52.

73. "Blinding Headlights," *Washington Post*, June 21, 1914, 14.

74. Clay McShane and Joel Tarr, "The Centrality of the Horse in the Nineteenth-Century American City," in *The Making of Urban America*, ed. Raymond A. Mohl, 2nd ed. (Wilmington, DE: SR Books, 1997), 107–108.

75. Charles Babbitt, *The Law Applied to Motor Vehicles: Citing All the Reported Cases Decided during the First Fifteen Years of the Use of Motor Vehicles*, 2nd ed. (Washington, DC: Arthur Blakemore, 1917), 1159.

76. C. E. T. Scarps, "Confusion Bound to Arise From New Headlight Law," *New York Tribune*, July 1, 1917, B8. See also E. L. Ferguson, "Motor Laws in Various States Are Compared," *New York Times*, Jan. 7, 1917, A17.

77. From *Jaquith v. Worden*, a case heard in Washington. Cited in Babbitt, *The Law Applied to Motor Vehicles*, 309.

78. "Automobiles. Care Required of Automobile Drivers. Pedestrians," *Virginia Law Review* 2, no. 4 (January 1915): 298–299.

79. McShane and Tarr, "The Centrality of the Horse in the Nineteenth-Century American City," 118.

80. For example, "Auto Headlight Glare Cases Up," *Christian Science Monitor*, Aug. 2, 1917, 9.

81. "Headlight Law Still Puzzles Automobilists," *Christian Science Monitor*, July 12, 1916, 1. See also the chart of regulations made to draw "lawmakers' attention to the wide divergency in the requirements of many States and municipalities" concerning automobile headlights. "Glaring Differences in Anti-Glare Laws," *Horseless Age* 37, no. 3 (Feb. 1, 1916): 146–147.

82. "Auto Lamps a Puzzle," *Washington Post*, Aug. 23, 1914, 12.

83. Reported in "Statutory Construction: Statute as to Automobile Lights Void for Indefiniteness," *Michigan Law Review* 18, no. 8 (June 1920): 810–811.

84. Replogle, "The Determination of Headlight Glare," at 890.

85. "Definition of Glare Needed For Headlight Law," *Christian Science Monitor*, Feb. 17, 1917, 9; "Control of Glare of Auto Lights Problem Still before Engineers," *Washington Post*, Apr. 22, 1917, 15. The most useful summary of headlight advances at the time is found in P. G. Nutting, "An Analysis of the Automobile Headlight Problem," *Journal of the Franklin Institute* 196, no. 4 (October 1923): 529–535.

86. E. E. Brakeman and C. S. Slocombe, "A Review of Recent Experimental Results Relevant to the Study of Individual Accident Susceptibility," *Psychological Bulletin* 26, no. 1 (January 1929), 16.

87. Clayton H. Sharp, "Safety Devices Necessary," *Motordom* 11 (May 1918): 13.

88. Percy Cobb, "Dark-Adaptation with Especial Reference to the Problems of Night Flying," *Psychological Review* 26, no. 6 (November 1919): 444; Leonard Troland, "Vision—General Phenomena," *Psychological Bulletin* 16, no. 4 (April 1919): 126.

89. "Light Law Is Drastic," *Washington Post*, Sept. 27, 1914, 12.

90. "Placing the Lights So As to Get Best Results from Them," *Christian Science Monitor*, July 13, 1912, 10.

91. "Lights for 1-2 Road Only Advocated," *New York Times*, Sept. 3, 1916, XX9; "Says Just Paint Bulbs to Meet Headlight Law," *New York Times*, Aug. 5, 1917, E4.

92. For example, "Dazzling by Motor-Car Lamps: Protective Devices," *British Medical Journal* 1, no. 3362 (June 6, 1925): 1046.

93. "Auto Headlight Cases Tried in Municipal Court," *Christian Science Monitor*, May 10, 1916, 5.

94. "Headlights a Danger."

95. The growing variety of automobile accessories had become a large part of manufacturers' business activity, including a range of headlamps, accessories, and tools to adjust them. John Jakle, *City Lights: Illuminating the American Night* (Baltimore, MD: Johns Hopkins University Press, 2001), 114; "Automobile Accessories Constitute Index of Active Business," *Christian Science Monitor*, Mar 6, 1909, 9.

96. Harold Whiting Slauson, "New Aids for the Motorist," *Outing Magazine* 57, no. 6 (March 1911): 758.

97. "Twelve Electric Bulbs on Wind Shield Give Satisfaction Without Blinding," *Washington Post*, July 26, 1914, M1. [First published in *Paris L'Illustration*.]

98. "Night Driving Made Safe by Cadillac," *Los Angeles Times*, Nov. 23, 1919, VI4.

99. A. R. Dennington, "Installation, Operation Power Application: A Record of Success," *Electrical Age* 49, no. 4 (Oct. 1, 1916): 37.

100. Ibid., 37–38.

101. Osgood Advertisement, *Chicago Daily Tribune*, Aug. 6, 1916, 8.

102. "Standard Glare Law," *Washington Post*, July 28, 1918, 18.

103. "Daredevil Motoring Thing of the Past," *Atlanta Constitution*, Aug. 6, 1916, A10. See also "Care Needed at Night," *Boston Daily Globe*, Mar. 6, 1910, 60, and "Road Etiquette," *Los Angeles Times*, Aug. 15, 1915, VI4.

104. See, for example, Lawrence Augustus Averill, "An Important Factor in Ocular Hygiene: Glare," *Mind and Body* 28, no. 303 (December 1921): 806.

105. Regardless of greater standardization, driving at night was not as safe as in the day. Accidents at night, when the overall volume of traffic was greatly reduced, nevertheless comprised 30 percent of the total number of accidents in 1917. Twenty years later, in 1936, 55 percent of total traffic accidents occurred at night. Fatalities were even worse. Nighttime rates more than doubled from 1921 to 1936, even as daytime rates remained relatively constant. Discussed in L. A. S. Wood, "Problems of Highway Lighting," *Electric Journal* 33, no. 10 (October 1936): 459. My thanks to Basie Gitlin for this reference. See also Jakle, *City Lights*, 116ff.

106. Lavinia Riker Davis, *Journals of Lavinia Riker Davis* (New York: Privately published, 1964), 22–23.

107. "Midnight Ride on a Cow-Catcher," *Oneida Circular*, Aug. 24, 1868, 181.

108. See Tom Gunning, "The Cinema of Attractions: Early Film, Its Spectator and the Avant-Garde," in *Early Cinema: Space, Frame, Narrative*, ed. Thomas Elsaesser and Adam Barker (London: British Film Institute, 1990), and "Before Documentary: Early Nonfiction Films and the 'View' Aesthetic," in Daan Hertogs and Nico de Klerk, eds., *Uncharted Territory: Essays on Early Nonfiction Film* (Amsterdam: Nederlands Filmmuseum, 1997).

109. Gibson and Crooks, "A Theoretical Field-Analysis of Automobile Driving," 456–457, 465.

110. "Care Needed At Night"; "Motor Headlights," *Hartford Courant*, Nov. 9, 1913, A8.

111. See Paul Virilio, "Dromoscopy, or the Ecstasy of Enormities," *Wide Angle* 20, no. 3 (July 1998): 14, and his *Negative Horizon: An Essay in Dromoscopy,* trans. Michael Degener (London:

Continuum, 2005). Richard Hawkins's challenge to address social and phenomenological effects of cars has been broadly taken up since he voiced it more than thirty years ago in Richard Hawkins, "A Road Not Taken: Sociology and the Neglect of the Automobile," *California Sociologist* 9, no. 1/2 (1986): 61–79.

112. In some setups, this meant creating two clearly defined black spots in the projection, which were the shadows of the filament holder, which might have its own seating device. Once defined, the driver would move the bulbs forward until the spots disappeared. See High Speed, "Safety, Decency, and Law Demand Headlight Care," *Chicago Daily Tribune*, Oct. 7, 1917, D8.

113. If the headlight rose above that level, the driver would need to go back to the car to adjust the headlamp assembly downward. See "Keeping the Light Close to the Road," *Chicago Daily Tribune*, Feb. 4, 1917, D6.

114. G. L. Sealey, "Recommendations for Focusing Automobile Headlamps," *Automotive Industries. The Automobile* 38, no. 8 (Feb. 21, 1918): 414.

115. Robert Slofs, "Equipping Your Automobile: Tools for the Real Troubles," *Outing Magazine* 55, no. 4 (January 1910): 452; High Speed, "Safety, Decency, and Law Demand Headlight Care," D8; "Headlight Adjusting Classes Given Free," *Los Angeles Times*, Aug. 8, 1925, A6.

116. E. A. Greathed, "Some Aspects of the Automobile," *Monthly Review* 19, no. 56 (May 1905): 59–60. Drivers kept a range of tools in the car and had to be prepared to use any of them at any time.

117. Gibson and Crooks, "A Theoretical Field-Analysis of Automobile Driving," 469–471.

118. "Care Needed at Night."

119. Sigfried Giedion, *Space, Time and Architecture: The Growth of a New Tradition* (Cambridge, MA: Harvard University Press, 1941), 729–730, 431–432.

120. Gibson and Crooks, "A Theoretical Field-Analysis of Automobile Driving," 458, 471.

CHAPTER 4: LIGHTING FOR LABOR

1. C. W. Price, "Light, the Key Tool of Industry," *American Industries* 21, no. 1 (August 1920): 18.

2. Cited in Roger Ekirch, *At Day's Close: Night in Times Past* (New York: Norton, 2005), 163.

3. Ekirch, *At Day's Close*, 155–156, 168–173.

4. Bryan D. Palmer, *Cultures of Darkness: Night Travels in the Histories of Transgression* (New York: Monthly Review Press, 2000), 139–142. Karl Marx discussed night trades in *Capital: A Critical Analysis of Capitalist Production*, ed. Friedrich Engels, trans. Samuel Moore and Edward Avelin (London: Swan Sonnenschein and Co., 1887), 1:249–250.

5. Ekirch, *At Day's Close*, 165–167.

6. Cited in E. P. Thompson, "Time, Work-Discipline, and Industrial Capitalism," *Past and Present* 38 (December 1967): 73. See also Ekirch, *At Day's Close*, 156–164, and Palmer, *Cultures of Darkness*, 143.

7. Ekirch, *At Day's Close*, 163–164, 182.

8. Gertrude Whiting, *A Lace Guide for Makers and Collectors* (New York: E. P. Dutton and Co., 1920), 28–29. See also Jane Brox, *Brilliant: The Evolution of Artificial Light* (Boston: Houghton Mifflin, 2010), 18–19; and Gertrude Whiting, *Tools and Toys of Stitchery* (New York: Columbia University Press, 1928).

9. Ekirch, *At Day's Close*, 164.

10. Maureen Dillon, *Artificial Sunshine: A Social History of Domestic Lighting* (London: National Trust, 2002), 39–40.

11. Thomas Decker, "Candlelight; Or, The Nocturnall Tryumph," in *The Seven Deadly Sins of London* (1606) (London: English Scholar's Library, 1879), 24.

12. Thomas Cooper, *Some Information Concerning Gas Lights* (Philadelphia: John Conrad and Co., 1816), 10.

13. Ekirch, *At Day's Close*, 162.

14. Brox, *Brilliant*, 17; Dillon, *Artificial Sunshine*, 42.

15. Dillon, *Artificial Sunshine*, 33; Brox, *Brilliant*, 63–65.

16. Allen Kiefer Gaetjens and Dean M. Warren, *Lighting for Production in the Factory* (Cleveland, OH: General Electric Co., NELA Park Engineering Dept., January 1939), inside front cover.

17. William Fairbairn, *Treatise on Mills and Millwork*, Part I, 3rd ed. (London: Longmans, Green, 1871), 1–10, and Part II, 2nd ed. (1863), 110–116. For more recent histories of the factory as a building type, see Nikolaus Pevsner, *A History of Building Types* (Princeton, NJ: Princeton University Press, 1976), 273–288; Robert B. Gordon and Patrick M. Malone, *The Texture of Industry: An Archaeological View of the Industrialization of North America* (New York: Oxford University Press, 1994); Lindy Biggs, *The Rational Factory: Architecture, Technology, and Work in America's Age of Mass Production* (Baltimore, MD: Johns Hopkins University Press, 1996); Betsy Bradley, *The Works: The Industrial Architecture of the United States* (Oxford: Oxford University Press, 1998); and Gillian Darley, *Factory* (London: Reaktion Books, 2003).

18. Daniel Nelson, *Managers and Workers, Origins of the Twentieth-Century Factory System in the United States 1880–1920* (Madison: University of Wisconsin Press, 1995), x. Fairbairn believed the mills' stratified workforce would become a model for all manufacturing. Fairbairn, *Treatise on Mills and Millwork*, Part I, 9.

19. "The Influence of the War on Industrial Lighting," *Electrical Review* 73, no. 20 (Nov. 16, 1918): 775.

20. William Pierson, Jr., "Notes on Early Industrial Architecture in England," *Journal of the Society of Architectural Historians*, 8, no. 1/2 (January–June 1949), 2.

21. Ibid., 6.

22. Warren D. Devine, Jr., "From Shafts to Wires: Historical Perspective on Electrification," *Journal of Economic History* 43, no. 2 (June 1983): 347–372.

23. Bradley, *The Works*, 162. Builders enhanced daylighting with pilasters to strengthen walls and widen windows or by splaying jambs, but these were incremental improvements. See Bradley, *The Works*, 229.

24. Cited in George M. Price, *The Modern Factory: Safety, Sanitation and Welfare* (New York: John Wiley and Sons, 1914), 235. Fairbairn had already noted this tendency: *Treatise on Mills and Millwork*, Part II, 72.

25. F. A. Patterson, "Factory Illumination," *American Architect* 107, no. 2044 (Feb. 24, 1915): 135–136.

26. Clarence Clewell, "Factory Lighting," *Architecture* 38, no. 3 (September 1918): 236. See also Bradley, *The Works*, chap. 7.

27. C. J. H. Woodbury, "Electric Lighting in Mills," in *Proceedings of the Semi-Annual Meeting of the Northeast Cotton Manufacturers' Association* 33 (Boston: Franklin Press and Rand, Avery, and Co., 1883), 18; Gordon and Malone, *The Texture of Industry*, 316.

28. Marx, *Capital*, 1:305.

29. Cited in Ian West, "Light Satanic Mills: The Impact of Artificial Lighting in Early Factories" (Ph.D. diss., University of Leicester, December 2008), 6.

30. M. E. Falkus, "The Early Development of the British Gas Industry, 1790–1815," *Economic History Review*, New Series, 35, no. 2 (May 1982): 219. See also Wolfgang Schivelbusch, *Lichtblicke: Zur Geschichte der Künstlichen Helligkeit im 19. Jahrhundert* (Munich: Karl Hanser, 1983), 16ff.

31. Peter Baldwin, *In the Watches of the Night: Life in the Nocturnal City, 1820–1930* (Chicago: University of Chicago Press, 2012), 36–37.

32. A mill owner encouraged Boulton and Watt, makers of steam engines, to develop gas lighting as a new business. Falkus, "The Early Development of the British Gas Industry," 222–223. See also Leslie Tomory, *Progressive Enlightenment: The Origins of the Gaslight Industry, 1780–1820* (Cambridge, MA: MIT Press, 2012), 69.

33. Cooper, *Some Information Concerning Gas Lights*, 12–13.

34. P. G. M. Dickson, *The Sun Insurance Office 1710–1960* (London: Oxford University Press, 1960), 92, 302; C. Trebilcock, *Phoenix Assurance and the Development of British Insurance*, vol. 1, *1782–1970* (Cambridge: Cambridge University Press, 1985), 354ff; William Murdoch, "An Account of the Application of the Gas from Coal to Economical Purposes," *Philosophical Transactions of the Royal Society of London* 98 (1808): 130. Frequent fires were also an impetus for mill owners to experiment with iron construction.

35. William Murdoch, a pioneer in the commercialization of gas light, granted that it was difficult in practice to equalize all the flames in gas lamps. Murdoch, "An Account of the Application of the Gas from Coal to Economical Purposes," 124.

36. Brox, *Brilliant*, 61. Makeshift photometers were deemed to be "accurate enough" for most purposes. Cooper, *Some Information Concerning Gas Lights*, 11–12; Baldwin, *In the Watches of the Night*, 16.

37. Edward Baines, *History of the Cotton Manufacture in Great Britain …* (London: H. Fisher, R. Fisher, and P. Jackson, 1835), 243–244. American developments followed similar lines; see Baldwin, *In the Watches of the Night*, 36–37; Marshall B. Davidson, "Early American Lighting," *Metropolitan Museum of Art Bulletin*, New Series, 3, no. 1 (Summer 1944): 30–40, at 39; and Emerson McMillin, "American Gas Interests," in *One Hundred Years of American Commerce, 1795–1895*, ed. Chauncey Depew (New York: D. O. Haynes and Co., 1895), 295.

38. Woodbury, "Electric Lighting in Mills," 17.

39. Baldwin, *In the Watches of the Night*, 37.

40. "Experience with Electric Light in Mills," in *Proceedings of the Semi-Annual Meeting of the Northeast Cotton Manufacturers' Association* 33 (Boston: Franklin Press and Rand, Avery, and Co., 1883), 27.

41. Regarding night workers, "compared with the countless millions who work by day, their numbers are of course infinitely small." Journeyman Engineer (Thomas Wright), *The Great Unwashed* (London: Tinsley Brothers, 1868), 188–190.

42. Baldwin, *In the Watches of the Night*, 37.

43. William Dodd, cited in James R. Simmons, Jr., ed., *Factory Lives: Four Nineteenth-Century Working-Class Autobiographies* (Ontario: Broadview Press, 2007), 241. Dodd's book was published in 1841 as *A Narrative of the Experience and Sufferings of William Dodd* (London: L. G. Seeley, 1841).

44. C. M. Goddard, "The Evolution of the 'Code,'" *Electrical World* 80, no. 11 (Sept. 9, 1922): 551–553, at 551.

45. Bradley, *The Works*, 106; David E. Nye, *Electrifying America: Social Meanings of a New Technology, 1880–1940* (Cambridge, MA: MIT Press, 1992), 191–192.

46. Maury Klein, *The Power Makers: Steam, Electricity, and the Men Who Invented Modern America* (New York: Bloomsbury Press, 2008), 113–117; "The Brush Electric Light," *Scientific American* 44, no. 14 (Apr. 2, 1881): 211.

47. Goddard, "The Evolution of the 'Code,'" 551.

48. Nye, *Electrifying America*, 191.

49. Woodbury, "Electric Lighting in Mills," 14. Woodbury's reports were widely reprinted in scientific and trade journals such as *Mechanics, Journal of the Franklin Institute, Journal of Engineering and Mechanical Progress, American Engineer*, and the like. "Woodbury, Charles Jeptha Hill," in *The National Cyclopedia of American Biography*, vol. 12, James Terry White, general ed. (New York: James T. White and Co., 1904), 81.

50. Allen Ripley Foote, *Economic Value of Electric Light and Power* (Cincinnati: Robert Clarke and Co., 1889), 21.

51. On the diverse routes to factory electrification, see Colum Giles and Ian Goodall, *Yorkshire Textile Mills: The Buildings of the Yorkshire Textile Industry 1770–1930* (London: HMSO, 1992): 163. See Daniel R. Shiman, "Explaining the Collapse of the British Electrical Supply Industry in the 1880s: Gas versus Electric Lighting Prices" *Business and Economic History* 22, no. 1 (Fall 1993): 318–327.

52. Mr. Ludlum, a plant manager, recalled that the first lamps in his mill were installed with uninsulated wires. "Experience with Electric Light in Mills," 47.

53. "Electric Lighting and Insurance," *Electric Age*, Apr. 26, 1890, 3.

54. William Brophy, "Relation of Insurance to Electric Lighting and Power," *Electrical Age* 11, no. 9 (Mar. 4, 1893): 137.

55. C. J. H. Woodbury, "The Relation of Electric Lighting to Insurance," in *Proceedings of the National Electric Light Association* 4 (Baltimore, MD: Baltimore Publishing, 1886): 188–190, at 189. The paper, read at the national conference, was widely reprinted.

56. C. J. H. Woodbury, "The Relation of Electric Lighting to Insurance," in *The Insurance Yearbook, 1886–1887* (New York: Spectator Co., 1886), 200–204, at 202.

57. Cited in Nye, *Electrifying America*, 190–191.

58. "Experience with Electric Light," 26, and Woodbury, "Electric Lighting in Mills," 14–15.

59. Woodbury, "Electric Lighting in Mills," 18.

60. Letter to the editor, *Iron Trade Review* 76 (April 1925): 1131.

61. Discussion following Woodbury, "Electric Lighting in Mills," 44.

62. "Experience with Electric Light," 28.

63. Stephen Greene, "Modifications in Mill Design Resulting from Changes in Motive Power," *Transactions of the New England Cotton Manufacturers' Association* 63 (1898): 128–136; Sidney B. Paine, *Development of the Electric Drive* (Schenectady, NY: General Electric Co., 1909); Nelson, *Managers and Workers*, 18.

64. Nye, *Electrifying America*, 199, 222–234. Other benefits of unit drives are described in Devine, "From Shafts to Wires," 350–356; J. Shaw, "Development of Textile Electric Driving," *Electrician*, Feb. 7, 1908, 632; F. B. Crocker, V. M. Benedict, and A. F. Ormsbee, "Electric Power In Factories And Mills," *Transactions of American Institute of Electrical Engineers* 12 (1896): 298–299; and Paine, *Development of the Electric Drive*.

65. Greene, "Modifications in Mill Design Resulting from Changes in Motive Power," 134.

66. Devine, "From Shafts to Wires," 363–364.

67. Biggs, *The Rational Factory*, 86–87.

68. For more on electrification's effects on factory design generally, see Nye, *Electrifying America*, 185–237, and Biggs, *The Rational Factory*, 90–94.

69. Devine, "From Shafts to Wires," 349–362; Nye, *Electrifying America*, 186–200; Biggs, "The Rational Factory. Highland Park's New Shop, 1914–1919," in *The Rational Factory*, 95–117; Ilene R. Tyler, "Highland Park Ford Plant: Documentation and Redevelopment," *APT Bulletin* 46, no. 2/3 (2015): 36–43.

70. E. B. Rowe and Frank B. Rae, Jr., "The Illumination of Mills and Factories with Small Units," *Electrical Review and Western Electrician* 55, no. 11 (Sept. 11, 1909): 503.

71. E. Leavenworth Elliott, "Efficiency," *Good Lighting and the Illuminating Engineer* 5, no. 10 (December 1910): 497.

72. Ward Harrison, "The Necessity for Standards in the Relation between Illumination and Output," *TIES (Transactions of the Illuminating Engineering Society)* 15, no. 5 (July 20, 1920): 337–339, at 338.

73. "Lighting Opportunities and Responsibilities," *Electrical Review* 73, no. 10 (Sept. 17, 1918): 369.

74. James Marsden Fitch, *American Building 2: Environmental Forces That Shape It* (New York: Schocken, 1975), 89, cited in Nye, *Electrifying America,* 191.

75. C. J. H. Woodbury, *The Engineer as an Economist* (Boston: C. J. H. Woodbury, 1905), 17.

76. Janice Gross Stein, *The Cult of Efficiency* (Toronto: Anansi, 2001), 20. Samuel Hays, *Conservation and the Gospel of Efficiency: The Progressive Conservation Moment, 1890–1920* (Cambridge, MA: Harvard University Press, 1959). For secondary sources on efficiency and scientific management in regard to industrial production, see Anson Rabinbach, *The Human Motor: Energy, Fatigue, and the Origins of Modernity* (New York: Basic Books, 1990); Judith Merkle Riley, *Management and Ideology: The Legacy of the International Scientific Movement* (Berkeley: University of California Press, 1980), 50; Alfred D. Chandler, *The Visible Hand: The Managerial Revolution in American Business* (Cambridge, MA: Harvard University Press, 1977); and Daniel Nelson, *Frederick W. Taylor and the Rise of Scientific Management* (Madison: University of Wisconsin Press, 1980).

77. Lewis Mumford, *The Myth of the Machine: The Pentagon of Power* (New York: Harcourt Brace Jovanovich, 1970), 172–173.

78. "Dr. Matthew Luckiesh Is Dead," *New York Times,* Nov. 3, 1967, 43.

79. Matthew Luckiesh, *Light and Work: A Discussion of Quality and Quantity of Light in Relation to Effective Vision and Efficient Work* (New York: Van Nostrand Co., 1924), 172–177.

80. Ibid., 199.

81. Ibid., 184.

82. Ibid., 208–209.

83. On speed as a facet of modernity, see Stephen Kern, "Speed," in *The Culture of Time and Space, 1880–1918* (Cambridge, MA: Harvard University Press, 2003), 109–130; Jeffrey Schnapp, "Crash (Speed as Engine of Individuation)," *Modernism/Modernity* 6, no. 1 (January 1999): 1–49; Wolfgang Schivelbusch, "Railroad Space and Railroad Time," *New German Critique* 14 (Spring 1978): 31–40; and Wolfgang Sachs, "Victorious Speed," in *For Love of the Automobile: Looking Back Into the History of Our Desires*, trans. Don Reneau (Berkeley: University of California Press, 1992), 110–124.

84. Luckiesh, *Light and Work*, 187. Luckiesh's key reference regarding the eye's efficiency is the work of Clarence E. Ferree. For a collection of writings by Ferree and his wife, Gertrude Rand, see *Contributions from the Psychological Laboratory*, Bryn Mawr College Monographs, Reprint Series, 12 (Bryn Mawr, PA: Bryn Mawr College, March 1922).

85. J. S. Dow, "The Relative Merits of Direct and Indirect Lighting," *Illuminating Engineer* 2, no. 8 (October 1907): 590.

86. For example, Luckiesh, *Light and Work*, 288–289.

87. Jonathan Crary, *Suspensions of Perception: Attention, Spectacle, and Modern Culture* (Cambridge, MA: MIT Press, 2001), 19.

88. John Dewey, *The Early Works, 1882–1898*, vol. 2, *1887: Psychology* (Carbondale: Southern Illinois University Press, 1975), 118–119; Crary, *Suspensions of Perception*, 24.

89. Michel Foucault, *Discipline and Punish: The Birth of the Prison* (New York: Vintage, 1977), 193.

90. William Durgin, "Productive Intensities," *TIES (Transactions of the Illuminating Engineering Society)* 13, no. 8 (Nov. 20, 1918): 418. Emphasis in original.

91. The economic value of light was considered well before electricity. English "laws of ancient lights" divided natural light into productive and nonproductive classes. Only the former could be legally protected. Electrification made the distinction irrelevant. Robert Kerr, *On Ancient Lights and the Evidence of Surveyors Thereon with Tables for the Measurement of Obstructions* (London: J. Murray, 1865); Homersham Cox, *The Law and Science of Ancient Lights* (London: H. Sweet, 1871); Alfred A. Hudson, *The Law of Light and Air* (London: Estates Gazette, 1905); and Alexandre Kedar, "The History of Anglo-American Legal Discourses about Obstruction of Lights" (Ph.D. diss., Harvard Law School, May 1989). See also David L. DiLaura, "Light's Measure: A History of Industrial Photometry to 1909," *LEUKOS: Journal of the Illuminating Engineering Society* 1, no. 3 (2005): 75–149.

92. Woodbury, "Electric Lighting in Mills," 18.

93. Discussion following Woodbury, "Electric Lighting in Mills," 44.

94. Ibid., 46–47.

95. Edward Atkinson, "Electric Lighting for Factories," in *Proceedings of the Semi-Annual Meeting of the New England Cotton Manufacturers' Association* 33 (Boston: Franklin Press and Rand, Avery, and Co., 1883), 49–58, at 51.

96. C. F. Scott, "Cost and Value of Light," *Electric Journal* 7, no. 5 (May 1910): 333–336. Scott may have been influenced by a more general argument regarding the setting of wages published the year before: Harrington Emerson, *Efficiency As a Basis for Operation and Wages* (New York: Engineering Magazine, 1909). On Scott's career, see Charles F. Scott, "The Young Engineer," *Electric Club Journal*, May 1904, 198–204; and "Guide to the Charles Felton Scott Papers," MS 779, Manuscripts and Archives, Yale University.

97. Scott, "Cost and Value of Light," 333–334.

98. "Recent Developments in Factory Lighting," *American Architect* 98, no. 1804 (July 20, 1910): 24.

99. "Changes in the Faculty," *Yale Scientific Monthly* 19, no. 1 (September 1912): 60–61.

100. Clarence E. Clewell, *Factory Lighting* (New York: McGraw-Hill, 1913), vi.

101. Ibid., 100.

102. Ibid., 1.

103. Ibid., 113.

104. Ibid., 100.

105. Ibid., 113–114.

106. See William Ennis, "The Relation of Purchasing to Production," *Engineering Magazine* 29, no. 4 (July 1905): 519–536, at 528. Durgin, "Productive Intensities," 418.

107. Editorial, "Good Results from Chicago Industrial Lighting Survey," *Electrical Review* 73, no. 18 (Nov. 2, 1918): 693.

108. Durgin, "Productive Intensities," 418.

109. C. E. Clewell, "Economic Aspects of Industrial Lighting," *Electrical World* 73, no. 8 (Feb. 22, 1919): 371–374.

110. James Cravath, "The Human Factor in Industrial Lighting," *Journal of Electricity and Western Industry* 46, no. 4 (Feb. 15, 1921): 181.

111. Geoffrey C. Bowker and Susan Leigh Star, *Sorting Things Out: Classification and Its Consequences* (Cambridge, MA: MIT Press, 1999), 13.

112. JoAnne Yates, *Control through Communication: The Rise of System in American Management* (Baltimore, MD: Johns Hopkins University Press, 1989), xv–xvii, 11–13.

113. Roger Henning, "Selling Standards," in *A World of Standards*, ed. Nils Brunsson and Bengt Jacobsson and Associates (New York: Oxford University Press, 2000), 114–124; and Nils Brunsson and Bengt Jacobsson and Associates, "Following Standards," in Brunsson and Jacobsson, eds., *A World of Standards*, 127–137. Bowker and Star, *Sorting Things Out*, 37–39. Practically every field developed standard procedures and products in the early 1900s. As secretary of commerce, Herbert Hoover wrote: "Standardization is the outstanding note of this century." *Standards Yearbook* (Washington, DC: National Bureau of Standards, 1927), 1. On the role of industry in setting standards, see David Noble, *America by Design: Science, Technology, and the Rise of Corporate Capitalism* (New York: Oxford University Press, 1979).

114. See, for example, Richard Henry Pierce and Robert E. Richardson, *The National Electrical Code: An Analysis and Explanation of the Underwriters' Electrical Code, Intelligible to Non-Experts* (Chicago: Charles Hewitt, 1896).

115. Goddard, "The Evolution of the 'Code.'"

116. Committee on Lighting Legislation and Committee on Factory Lighting, "Code of Lighting—Factories, Mills, and Other Work Places," *TIES (Transactions of the Illuminating Engineering Society)* 10, no. 8 (Nov. 20, 1915): 605–641, at 617–618. Clewell commented on the code's new emphasis on daylight, noting that it had been neglected in the past in favor of electric light because of its greater complexity: C. E. Clewell, "Practical Aspects of the Illuminating Engineering Society's New Code of Lighting," *Lighting Journal*, October 1915, 222; C. E. Clewell, "Establishing a Basis for Laws on Lighting," *Electrical World* 66, no. 26 (Dec. 25, 1915): 1417–1418.

117. Luckiesh, *Work and Lighting*, 176–177.

118. Elliott, "Efficiency," 497.

119. F .H. Bernhard, "Lighting of Rubber-Goods Factories," *Electrical Review* 73, no. 17 (Oct. 26, 1918): 56.

120. Luckiesh, *Work and Lighting*, 200.

121. "Factory Lighting," *British Medical Journal* 2, no. 2866 (Dec. 4, 1915): 830.

122. Friedrich Engels, *The Condition of the Working Class in England in 1844* (New York: Cosimo, 2008), 191.

123. Henry Mayhew, *Mayhew's London* (London: Spring Books, 1851), 167–168. Cited in Baldwin, *In the Watches of the Night*, 37. See also "The Use and Abuse of Eyesight," *Methodist Quarterly*

Review 43 (January 1861): 104–126. Margaret, a seamstress, agonized over approaching blindness in Elizabeth Gaskell, *Mary Barton: A Tale of Manchester Life* (Leipzig: Bernhard Tauchnitz, 1849). My thanks to Siobhan Carroll for these references. Mayhew, an English labor reformer, thought gas was more "pernicious" than candle light because of its glare and heat. He recalled his "eyes used to feel like two bits of burning coals in my head" when he had to work late by gaslight. Mayhew, *Mayhew's London*, 343.

124. Ted Underwood, *Literature, Science, and Political Economy, 1760–1860* (New York: Palgrave, 2005), 9.

125. Andreas Killen, *Berlin Electropolis: Shock, Nerves, and German Modernity* (Berkeley: University of California Press, 2006), 30–31.

126. Rabinbach, *The Human Motor*, 62.

127. Karl Marx believed that fatigue was a by-product of industrialization, the "premature exhaustion" of labor on the part of capital: *Capital*, 1:250. Hugo Münsterberg argued that fatigue was the result of machine speeds that disrupted the "subjective rhythmic experience" of machine operators: Hugo Münsterberg, *Psychology and Industrial Efficiency* (Boston: Houghton Mifflin, 1913), 98.

128. H. M. Vernon, *Industrial Fatigue and Efficiency* (London: George Routledge and Sons; New York: E. P. Dutton, 1921), 3. See also, A. J. McIvor, "Employers, the government, and industrial fatigue in Britain, 1890–1918," *British Journal of Industrial Medicine* 44 (1987): 724–732.

129. See Josephine Goldmark, *On Conserving the Nation's Energy: Fatigue and Efficiency. A Study in Industry* (New York: Russell Sage Foundation, 1912), 286.

130. Frank B. Gilbreth, *Motion Study: A Method for Increasing the Efficiency of the Workman* (New York: Van Nostrand Co., 1911), 50–51.

131. Ibid., 50–52.

132. F. B. Allen, "Important Considerations in Factory Lighting," *Electrical Review and Western Electrician* 59 (Aug. 12, 1911): 323.

133. B. H. Thwaite, *Our Factories, Workshops and Warehouses: Their Sanitary and Fire-Resisting Arrangements* (New York: E. and F. N. Spon, 1882), 112.

134. Atkinson, "Electric Lighting for Factories," 52.

135. G. Stanley Hall and Theodate L. Smith, "Reactions to Light and Darkness," *American Journal of Psychology* 14, no.1 (January 1903): 71.

136. For example, see Clewell, "Economic Aspects of Industrial Lighting," 1919, 373.

137. "Experience with Electric Lighting," 27, and discussion following Woodbury, "Electric Lighting in Mills," 36.

138. Rowe and Rae, "The Illumination of Mills and Factories with Small Units," 503.

139. J. M. Smith, "Practical Considerations in Cotton-Mill Illumination," *Electrical Review and Western Electrician* 59, no. 5 (Oct. 7, 1911): 739–740.

140. C. E. Clewell, "The Problem of Artificial Lighting: How to Evaluate Lighting Costs as a Proportion of Wages," *Industrial Management* 55, no. 1 (January 1918): 18. The presumption that better lighting spontaneously led to greater production was the basis for the Hawthorne experiments, beginning in 1924, which went on to become a case study in sociological studies of the effect of implicit bias. In this case, greater productivity turned out to be the result of workers' awareness that they were being observed for their response to brighter lighting. Although the premise of the experiments was falsified, it nonetheless demonstrates how inter-

nalized presumptions regarding lighting standards had become. See Elton Mayo, *The Human Problems of an Industrial Civilization* (New York: Macmillan, 1933); and Richard Gillespie, *Manufacturing Knowledge: A History of the Hawthorne Experiments* (New York: Cambridge University Press, 1991).

141. Raymond Williams, "Hegemony," *Marxism and Literature* (Oxford: Oxford University Press, 1977), 109–114.

142. Thompson, "Time, Work-Discipline, and Industrial Capitalism," 75.

143. Leon Gaster, "Ten Years of Illuminating Engineering: Its Lessons and Future Prospects," *Illuminating Engineer* 11, no. 1 (January 1918): 18.

144. Advertising "labor-saving" appliances, electrical equipment makers had long treated the home as a workplace. By 1930, electric utilities had established hundreds of dedicated home service departments that recommended ways to expedite further expansion of service to residences, usually with the acquisition of more lighting. Carolyn Goldstein, "From Service to Sales: Home Economics in Light and Power, 1920–1940," *Technology and Culture* 38, no. 1 (January 1997): 121–152.

145. Reginald C. Augustine, "School Illumination and Vision of Students," *American School Board Journal* 64, no. 2 (February 1922): 49–50.

146. Ellwood Cubberley, *Public School Administration* (New York: Houghton Mifflin 1916), 338. See also Raymond E. Callahan, *Education and the Cult of Efficiency: A Study of the Social Forces That Have Shaped the Administration of the Public Schools* (Chicago: University of Chicago Press, 1964).

147. "The Teaching of Hygiene in Schools. VI. Eye-Strain and Brain-Strain," *British Medical Journal* 1, no. 2299 (Jan. 21, 1905): 153–155, at 153–154.

148. Ada M. Hughes, "Relationship of the Kindergarten to Child Study," *Kindergarten Review* 5, no. 10 (June 1897): 507–508; A. K. Whitcomb, "Physical Defects of Children" (1899), in "Medical Inspection in the Public Schools," Joseph Lee and Margaret Curtis, eds., *Leaflets*, Massachusetts Civic League 7 (1906): 37; Roswell S. Wilcox, "Practical Hygiene in the Public Schools," *Medical Record* 66, no. 12 (Sept. 17, 1904): 455–458; Louis Bell, *The Art of Illumination*, 2nd ed. (New York: McGraw-Hill, 1912), 344.

149. J. Herbert Parsons, "Discussion on the Hygiene of Reading and Near Vision," *British Medical Journal* 2, no. 2799 (Aug. 22, 1914): 359. See also Casey A. Wood, "The Sanitary Regulation of the School-Room with Reference to Vision," *Elementary School Teacher* 7, no. 2 (October 1906): 62–71.

150. Augustine, "School Illumination and Vision of Students," 49–50.

151. The primary hazard was glare, the unhappy outcome of poorly situated desks, glossy or reflective interior finishes, misplaced electric lamps, and misguided attempts to brighten classrooms to meet factory standards. James Kerr, "Glare and the Student's Life," *Mind and Body* 26, no. 282 (November 1919): 241. Originally published in *School Hygiene* 7 (February 1916): 11–20.

152. "Teaching of Hygiene," 153.

153. Augustine, "School Illumination and Vision of Students," 49.

154. See appendix 2 in J. H. Berkowitz, *The Eyesight of School Children* (Washington, DC: Department of the Interior, Bureau of Education, 1919), 68–71.

155. Wisconsin, a leader in lighting regulations, used its factory lighting code as the basis of its 1918 school lighting code, which was issued by the State Industrial Commission. Before this, school lighting was treated as a brief section of the state's general Building Code. See "Proposed

School-Lighting Code for Wisconsin," *Electrical Review* 77, no. 20 (Nov. 13, 1920): 753–754; "Lighting Code for Factory and School," *Safety* 9, no. 4 (April 1922): 101–102; and Mark Rose, *Cities of Light and Heat: Domesticating Gas and Electricity in Urban America* (University Park: Pennsylvania State University Press, 1995), 98–106.

156. "Dark Days Ahead" (Washington, DC: Housekeepers' Chat Radio Service, U.S. Department of Agriculture Office of Information and Extension Service, Oct. 28, 1937), 1.

157. Ibid.

158. Claire Winslow, "'Better Light, Better Sight,' Is New Slogan," *Chicago Daily Tribune*, Sept. 29, 1935; "Eyesight Aids for Building More Popular: Painting and Decorating Stressed as Important to Lighting," *Washington Post*, Sept. 8, 1935; "The First Five Years," *Electrical World* 109 (1938): 48; "The Better Light–Better Sight Program," *Edison Electric Institute Bulletin*, January 1939, 7; "Remodeling the Lighting System," *Homemaker News*, no. 454 (Washington, DC: U.S. Department of Agriculture Press Service, Office of Information, and Extension Service, Oct. 4, 1940); "Ten Years of Better Light–Better Sight," *Electrical World* 122 (1944): 92. See also Margaret Maile Petty, "Threats and Promises: The Marketing and Promotion of Electric Lighting to Women in the United States, 1880s–1960s," *West 86th* 21, no. 1 (Spring-Summer 2014): 3–36; and Carolyn Goldstein, *Creating Consumers: Home Economists in Twentieth-century America* (Chapel Hill: University of North Carolina Press, 2012), 234–236.

159. M. Luckiesh and F. K. Moss, "The New Science of Seeing," *TIES (Transactions of the Illuminating Engineering Society)* 25 (1930): 15–49, at 15–18.

160. Review of Matthew Luckiesh and Frank K. Moss, *Journal of the American Medical Association* 110, no. 2 (Jan. 8, 1938): 152.

161. *See Your Home in a New Light: Tested Light-Conditioning Recipes That Create Light for Living* (Cleveland, OH: General Electric Co., Lamp Division, 1955). See also Margaret Maile Petty, "In Fear of Shadows: Light Conditioning the Postwar American Home and Lifestyle," in *Le Jeu Savant: Light and Darkness in 20th Century Architecture*, ed. Silvia Berselli et al. (Milan: Silvana, 2014), 259–268; and Margaret Maile Petty, "Cultures of Light: Electric Light in the United States, 1890s–1950s" (Ph.D. diss., Victoria University of Wellington, 2016), 172–180.

162. Advertisement, General Electric, *Popular Science* 131, no. 5 (November 1937): 7.

163. Hubert M. Langlois, "Public Utility Adjustment to Postwar Conditions," *Annals of the American Academy of Political and Social Science* 222 (July 1942): 152–155, at 153–154.

164. Arthur Rubin, ed., *Lighting Issues in the 1980s* (Washington, DC: National Bureau of Standards, U.S. Department of Commerce, July 1980), 96–97.

165. John Appel and James MacKenzie, "How Much Light Do We Really Need?," *Bulletin of the Atomic Scientists* 30, no. 10 (December 1974): 19.

166. Rubin, *Lighting Issues*, 96–97. See, for example, Thomas Ehrich, "Too Much Light? Critics Say Illumination in Many Buildings Is Greater Than Needed," *Wall Street Journal*, Sept. 28, 1972, 1; Donald K. Ross, *A Public Citizen's Action Manual* (New York: Grossman Publishing, 1973); Barry Keating, "Industry Standards and Consumer Welfare," *Journal of Consumer Affairs* 14, no. 2 (Winter 1980): 471–482; and Malcolm Spector and John I. Kitsuse, *Constructing Social Problems* (New Brunswick, NJ: Transaction Publishers, 2009), 146–147.

167. Appel and MacKenzie, "How Much Light Do We Really Need?," 18–24.

168. "Government Office Proves 'See Better–Work Better' Truth," *See Better–Work Better Bulletin* 3 (Cleveland, OH: General Electric Co., Lamp Division, 1959), 4–5.

CHAPTER 5: ELECTRIC SPEECH IN THE CITY

1. SMILAX, "Electric Light Advertising," *Electrical Magazine* 3, no. 6 (June 27, 1905): 511.

2. William R. Taylor, "Introduction," in *Inventing Times Square: Commerce and Culture of the Crossroads of the World*, ed. William R. Taylor (Baltimore, MD: Johns Hopkins University Press, 1996), xxv. Electric lighting was an explicit theme in other environments, such as international expositions (from the Third Exposition Universelle, held in Paris in 1878, when arc lighting was installed around the Paris Opera, to the 1937 Exposition Internationale, also in Paris, where Raoul Dufy's *La Fée électricité* (The Spirit of Electricity) impressed visitors to the Pavilion of Electricity and Light) and amusement parks, such as Blackpool, England, from 1879, and Luna Park, New York, from 1903. Luna Park was often compared to Times Square. On expositions, see David E. Nye, "Electrifying Expositions, 1880–1937," in *Narratives and Spaces* (New York: Columbia University Press, 1997). On Blackpool, see Tim Edensor and Steve Millington, "Blackpool," in *Cities of Light: Two Centuries of Urban Illumination*, ed. Sandy Isenstadt, Margaret Maile Petty, and Dietrich Neumann (New York: Routledge, 2015), 58–66. On Luna Park, see John Kasson, *Amusing the Millions: Coney Island at the Turn of the Century* (New York: Hill and Wang, 1978).

3. Marshall Berman, *On the Town: 100 Years of Spectacle in Times Square* (New York: Random House, 2006), 125.

4. Ibid., 126.

5. "Watts in a Name," *Edison Monthly* 13, no. 4 (April 1920): 68–70, at 69.

6. Cited in Bayrd Still, *Mirror for Gotham: New York as Seen by Contemporaries from Dutch Days to the Present* (New York: New York University Press, 1956), 188.

7. For more on city signs, see David Henkin, *City Reading: Written Words and Public Spaces in Antebellum New York* (New York: Columbia University Press, 1998); Sara Thornton, *Advertising, Subjectivity and the Nineteenth-Century Novel* (New York: Palgrave Macmillan, 2009); Maxine Berg and Helen Clifford, "Commerce and the Commodity: Graphic Display and Selling New Consumer Goods in Eighteenth-Century England," in *Art Markets in Europe, 1400–1800*, ed. M. North and D. Ormrod (New York: Routledge, 2016), 187–200; Stefan Haas, "Visual Discourse and the Metropolis: Mental Models of Cities and the Emergence of Commercial Advertising," in *Advertising and the European City: Historical Perspectives*, ed. Clemens Wischermann and Elliott Shore (Aldershot: Ashgate, 2000), 54–78; and "Street Signs Date from Ancient Days," *New York Times*, Jan. 27, 1929, W18.

8. These were the London Sky Signs Act of 1891 and the more comprehensive London Building Act of 1894. See Aaron J. Segal, "Commercial Immanence: The Poster and Urban Territory in Nineteenth-Century France," in Wischermann and Shore, eds., *Advertising and the European City*, 113–138. See also Clarence Moran, *The Business of Advertising* (Oxon: Routledge, 1905), 162–166. London's lighted advertisements were described in 1898 as "an infectious disease" in "The Man about Town," *Country Gentleman* 36, no. 1888 (July 16, 1898): 917.

9. Mary C. Henderson, *The City and the Theater: New York Playhouses from Bowling Green to Times Square* (Clifton, NJ: James T. White and Co., 1973).

10. Irving Lewis Allen, *The City in Slang: New York Life and Popular Speech* (New York: Oxford University Press, 1993), 59.

11. Henderson, *The City and the Theater*, 121.

12. "The Brush Electric Light," *Scientific American* 44, no. 14 (Apr. 2, 1881): 211–212; James McCabe, *New York by Sunlight and Gaslight* (Philadelphia: Douglas Brothers, 1882), 427; Ken Bloom, *Broadway: Its History, People, and Places. An Encyclopedia* (New York: Routledge, 2004), 490.

13. Ernest Ingersoll, *A Week in New York* (New York: Rand, McNally and Co., 1891), 139.

14. Stephen Burge Johnson, *The Roof Gardens of Broadway Theaters, 1883–1942* (Ann Arbor: University of Michigan Press, 1985), 7–13.

15. "Up under the Clouds," *New York Times*, May 27, 1892, 4. See also Vance Thompson, "The Roof-Gardens of New York," *Cosmopolitan* 27, no. 5 (September 1899): 503–514, esp. 506; Paul Van Du Zee, "New York Roof Gardens," *Godey's Magazine* 129, no. 770 (August 1894): 201–210, esp. 206; and Johnson, *Roof Gardens*, 17–21.

16. Cited in Allen, *The City in Slang*, 60.

17. Cited in Still, *Mirror for Gotham*, 260.

18. Rupert Hughes, *The Real New York* (London: Smart Set Publishing, 1904), 266. See also Valentine Cook, Jr., "The Great White Way," *Illuminating Engineer* 50, no. 3 (May 1906): 147.

19. Martin Banham, *The Cambridge Guide to Theater* (Cambridge: Cambridge University Press, 1998), 787–788.

20. "The Man Who Lights Up the Great White Way," *Signs of the Times* 20, no. 85 (May 1913): 14, 34, 37; Westbrook Pegler, "Man Who Turned Broadway at Night into Daylight Dies," *Atlanta Constitution*, Aug. 16, 1925; James Traub, *The Devil's Playground: A Century of Pleasure and Profit in Times Square* (New York: Random House, 2005), 41–46; Darcy Tell, *Times Square Spectacular: Lighting Up Broadway* (New York: Smithsonian Books, 2007).

21. The term's origins remain hazy. The nineteenth-century phrase "white way" signified a virtuous, even a holy path, as in Robert Williams Buchanan, "Retrospect: The Journey," in *White Rose and Red: A Love Story* (London: Strahan and Co., 1873), 170, or Rosamond Marriott Watson, "The White Way," *Independent* 51, no. 2643 (July 27, 1899): 1998. It could also carry racial implications, as in Nora Perry, "Major Molly's Christmas Promise," *Harper's Young People*, Dec. 20, 1892, 138. It might have first referred to lower Broadway following a heavy snowfall, alluding to a 1901 book titled *The Great White Way*, about a journey to Antarctica. See Allen, *The City in Slang*, 58–62. The most complete account is Barry Popik; see http://www.barrypopik.com/index.php/new_york_city/entry/great_white_way, July 8, 2016. See also Bloom, *Broadway*, 499.

22. Samuel G. Blythe, "The Itch for Publicity," *Saturday Evening Post* 181, no. 34 (Feb. 20, 1909): 8.

23. E. L. Elliott, "In Lightest New York," *Illuminating Engineer* 6, no. 4 (June 1911): 195.

24. Harry Tipper et al., *Advertising: Its Principles and Practice* (New York: Ronald Press, 1919), 489–493.

25. Walter Winchell, "On Broadway," *Daily Mirror*, 1935; Douglas Leigh, "This Is Times Square?," Sept. 19, 1964, Douglas Leigh Papers, Archives of American Art, Smithsonian Institution, series 7, reel 5847, p. 901 (hereafter cited as Leigh MSS).

26. Quoted in Bloom, *Broadway*, 289. "The Great Illuminator," *Lighting Dimensions* 7, no. 8 (1985), Leigh MSS, series 7, reel 5847, p. 660. Leigh went on to light skyscrapers and create visual decorative programs for political party conventions and the 1976 bicentennial. He won numerous awards and citations, including the key to New York City.

27. Frank C. Reilly, "The Newest in Electric Signs," *Good Lighting and the Illuminating Engineer* 7, no. 5 (July 1912): 247–249, at 249.

28. Will Irwin, *Highlights of Manhattan* (New York: Century Co., 1927), 325; William Leach, *Land of Desire: Merchants, Power, and the Rise of a New American Culture* (New York: Vintage, 1993), 342–343.

29. Dr. F. W. Gerhard Schmidt, "Illuminated Advertising," *Signs of the Times* 80, no. 3 (July, 1935): 48–50.

30. Leonard G. Shepard, "Sign Lighting," in Illuminating Engineering Society, *Lectures on Illuminating Engineering Practice* (New York: McGraw-Hill, 1917), 535–536.

31. "Austin Corbin Dead," *New York Times*, June 5, 1896, 1; "The Best 'Ad' in the City," *New York Times*, Oct. 4, 1896, 8.

32. Theodore Dreiser, cited in *Writing New York: A Literary Anthology*, ed. Phillip Lopate (New York: Washington Square Press, 1998), 340.

33. "Triumphant Industries: The General Electric Company," *Forum* 14 (Feb. 18, 1893): 7–8; Tell, *Times Square Spectacular*, 41–46; Traub, *The Devil's Playground*, 44–45.

34. Use of the term to denote an automatic electrical device begins to appear regularly in industry periodicals at the end of the first decade of the twentieth century. The earliest citation in the *OED* is from 1909. Shepard, "Sign Lighting," 535–537. The Reynolds Dull Company remained a leader in flasher manufacturing, adapting their product lines to the heavier loads of larger signs and for new lamp types such as neon. J. Willy Phelps, "The Origin and Development of the Motorless Flasher," *Signs of the Times* 27, no. 3 (November 1914); O. D. Ziegler, "Something about the Growth of the Flasher Industry," *Signs of the Times* 29, no. 1 (May 1919): 22–23; "Forty Years of Flasher History" *Signs of the Times* 74, no. 1 (May 1933): 15–16.

35. U.S. Patent 1,050,203, filed July 15, 1911, and granted January 14, 1913. Bickley's invention was not related to Thomas Edison's development of the Electro-Motograph, an enhancement for telephone receivers, although it was related to Edison's Automatic Telegraph System, which made use of perforated strips of paper for the transmission of messages. See "The Electro-Motograph," in Gaston Tissandier, *Popular Scientific Recreations* (New York: W. H. Stelle and Co., 1883), 262–264.

36. "The Talking Sign," *Electrical Review and Western Electrician* 59, no. 3 (July 15, 1911): 140.

37. Reilly, "The Newest in Electric Signs," 249.

38. The "Bickley Motograph or Continuous Talking Sign," *Electrical Review and Western Electrician* 561, no. 23 (Dec. 7, 1912): 1094; "Bickley Motograph," *Electrical Record* 12, no. 1 (July 1912): 50–51.

39. "The Newest Thing in Electrical Advertising," *Signs of the Times* 16, no. 76 (August 1912): 26.

40. "Bickley Motograph or Continuous Talking Sign," 1094; "Bickley Motograph," 50–51.

41. Ziegler, "Something about the Growth of the Flasher Industry," 22–23.

42. "Bickley Motograph or Continuous Talking Sign," 1094.

43. Reilly, "The Newest in Electric Signs," 247.

44. U.S. Patent 1,119,371, issued December 1, 1914. Reilly, "The Newest in Electric Signs," 248–249. See also "The Newest Thing in Electrical Advertising," 26. Through regular contact with theater figures, Reilly began writing plays and became a producer. "Writing on the Wall," *New Yorker*, Jan. 12, 1929, 11–12.

45. Lynn Sagalyn, *Times Square Roulette: Remaking the City Icon* (Cambridge, MA: MIT Press, 2003), 40; Reilly, "The Newest in Electric Signs," 249; "Huge Times Sign Will Flash News," *New York Times*, Nov. 8, 1928, 30; Meyer Berger, *The Story of The New York Times, 1851–1951* (New York: Simon and Schuster, 1951), 523–524.

46. "Leigh-Epok," *Tide*, Aug. 15, 1937, 28; Leigh MSS, series 7, reel 5841, p. 442.

47. Odette Keun, *I Think Aloud in America* (New York: Longmans, Green, 1939), cited in Still, *Mirror for Gotham*, 323.

48. Frank Ward, "ANIMATE. ... To Fascinate!" *Signs of the Times* 89, no. 2 (June 1938): 14. Although focused on animated puppets, the article's logic duplicates arguments for animated electric signs.

49. "Moving Signs along Broadway," *Edison Monthly* 4, no. 7 (December 1911): 225–227.

50. Advertisement, Wrigley's Chewing Gum, *New York Herald*, Apr. 29, 1920, 6.

51. By 1910 sign makers had at their disposal countless flasher effects, including "crawling snakes, jumping rabbits" and "fountains, streams of liquids, foam, smoke, fire, steam, cloud effect, revolving wheels, etc.," as well as various means to combine words and images. "Electric Sign Flashers," *Electrical Review and Western Electrician* 57, no. 24 (Dec. 10, 1910): 1210; "Moving Signs along Broadway," 225–227; Tama Starr and Edward Hayman, *Signs and Wonders: The Spectacular Marketing of America* (New York: Doubleday, 1998), 63–67; Traub, *The Devil's Playground*, 47; John Jakle, *City Lights: Illuminating the American Night* (Baltimore, MD: Johns Hopkins University Press, 2001), 201–202. Walter Benjamin noted that nineteenth-century promotions routinely joined imagery with text and were seen as being more effective for doing so. Walter Benjamin, *The Arcades Project*, trans. Howard Eiland and Kevin McLaughlin (Cambridge, MA: Harvard University Press, 1999), 689.

52. Francis Arthur Jones, "The Most Wonderful Electric Sign in the World, and How It Is Worked," *Strand Magazine* 42, no. 250 (October 1911): 443–448.

53. Max Horkheimer and Theodor W. Adorno, "The Culture Industry: Enlightenment as Mass Deception" (1944), in *Dialectic of Enlightenment: Philosophical Fragments*, ed. Gunzelin Schmid Noerr, trans. Edmund Jephcott (Stanford: Stanford University Press, 2002), 45–46, 94–136.

54. "Animated Sign Displays," *Signs of the Times* 77, no. 1 (May 1934): 10–11. The idea pertained to all displays, whether signs or show windows.

55. "Leigh's Biggest," *New Yorker*, Aug. 7, 1937, 10.

56. "Animation over Broadway: Douglas Leigh and Otto Messmer," program (New York: Museum of Modern Art, Feb. 7, 1993), organized by William Lorenzo and introduced by Douglas Leigh; "Program Announcement," Billboard Project Files, Leigh MSS, series 10, p. 40. Douglas Martin, "Douglas Leigh, the Man Who Lit Up Broadway, Dies at 92," *New York Times*, Dec. 16, 1999; Christopher Gray, "The Man behind Times Square's Smoke Rings," *New York Times*, Oct. 25, 1998.

57. "Animated Antics Make 'Em Glad, *Signs of the Times* 86, no. 3 (June 1937): 13.

58. "Leigh's Biggest," 10; Billboard Project Files, Leigh MSS, series 10, p. 56.

59. Robert Sellmer, "Douglas Leigh," *Life Magazine* 20, no. 13 (Apr. 1, 1946): 47–51. See also Robert Rochlin, "Bringing Advertising to Life," *Cornell Engineer* 7, no. 1 (October 1941): 10–11, 22; Arthur Liebers, "Animated Lighting to Sell the Theatre," *BoxOffice*, Jan. 25, 1947, Leigh MSS, series 9, General Scrapbooks, p. 1084.

60. Cited in Charles Musser, *Politicking and Emergent Media: US Presidential Elections of the 1890s* (Oakland: University of California Press, 2016), 74.

61. Ibid., 123–127.

62. Stephen Graham, *New York Nights* (New York: George H. Doran, 1927), 17.

63. Stephen Chalmers, "How Broadway's Magic Affects the Stranger: Impressions of a Scotsman," *New York Times*, Dec. 11, 1904, SM5.

64. O. J. Gude, "Art and Advertising Joined by Electricity," *Signs of the Times* 19, no. 79 (November 1912): 3.

65. Jones, "The Most Wonderful Electric Sign in the World," 443–444.

66. Nelson Fraser, *America: Old and New. Impressions of Six Months in the States* (London: John Ouseley, 1912), 119–120.

67. "Animated Antics Make 'Em Glad," 13.

68. "Katie Smith Speaks," typescript, June 12, 1945, Leigh MSS, Scrapbook 9, General Scrapbooks, 1943–48, p. 807.

69. Horkheimer and Adorno, "The Culture Industry," 99.

70. "A Plan for a White Way," *Dry Goods Reporter* 46, no. 1 (Mar. 20, 1915): 9.

71. John E. Roberts, "Advertising Value of Animated Electric Signs," *Public Service* 9, no. 4 (October 1910): 124. Trade articles emphasized that flashers reduced electrical loads by breaking the circuit repeatedly. "Electric Sign Flashers," *Electrical Review and Wester Electrician* 57, no. 24 (Dec. 10, 1910): 1210.

72. Ziegler, "Something about the Growth of the Flasher Industry," 22–23.

73. The Scintillator, "Electric Signs an Essential of Modern Business: Why They Prove So Effective," *Signs of the Times* 30, no. 4 (October 1919): 48, 51.

74. Tell, *Times Square Spectacular*, 41–46; Traub, *The Devil's Playground*, 44–45.

75. Reilly, "The Newest in Electric Signs," 248.

76. Ibid., 249.

77. Hugo Münsterberg, *Psychology and Industrial Efficiency* (Boston: Houghton Mifflin, 1913), 257–263.

78. Ibid., 281.

79. Henry Foster Adams, *Advertising and Its Mental Laws* (New York: Macmillan, 1916), 326–328.

80. Ibid., 51.

81. Tipper et al., *Advertising*, 83. See also Edward K. Strong, Jr., "Psychological Methods as Applied to Advertising," *Journal of Educational Psychology* 4 (1913): 393–404; George Oilar, "Points on the Psychology of Attention," *Business Philosopher* 7 (January 1911): 31–35; and Matthew Luckiesh, "Attention-Value," *Light and Color in Advertising and Merchandising* (New York: Van Nostrand, 1923), 72–86.

82. Janet Mabie, "Signs of the Times," *Christian Science Monitor*, Sept. 22, 1937, WM6.

83. Douglas Leigh, "This Business of Selling Big Spectaculars," *Signs of the Times* 78, no. 4 (December 1934): 18.

84. Ibid.

85. "Animated Sign Displays," *Signs of the Times* 77, no. 1 (May 1934): 10.

86. "Streamlined Advertising," *Signs of the Times* 80, no. 2 (June 1935): 41.

87. Keun, *I Think Aloud in America*, cited in Still, *Mirror for Gotham*, 323.

88. Guilds' Committee for Federal Writers' Publications, "Times Square District," in *New York City Guide* (New York: Random House, 1939), 171.

89. "Some Factors in the Use of Spectacular Electric Signs and an Estimate of Their Value," *Signs of the Times* 34, no. 3 (March 1920): 36.

90. Gude, "Art and Advertising Joined by Electricity," 3. Years later, noting the forceful splintering of his concentration in a shopping arcade, Walter Benjamin described the effect: "One's attention is spirited away as though by violence, and one has no choice but to stand there and remain looking up until it returns." Benjamin, *Arcades Project*, 60–61.

91. From a 1907 *New York Times* editorial, cited in Eric Gordon, *The Urban Spectator: American Concept Cities from Kodak to Google* (Lebanon, NH: Dartmouth College Press, 2010), 79.

92. Platon Michailowitsch Kerschenzew, *Das Schöpferische Theater* (Hamburg: Verlag Carl Hoym Nachf., 1922), 2. Kerschenzew suggested that Times Square in general was "a bacchanal of theater," p. 2.

93. Arnold Bennett, *Your United States: Impressions of a First Visit* (New York: George H. Doran, 1912), 21.

94. Rupert Hughes, *What Will People Say?* (New York: Harper and Bros., 1914), 16.

95. Rupert Brooke, *Letters from America* (New York: Charles Scribner's Sons, 1919), 30–34.

96. Chesterton, *What I Saw in America* (New York: Dodd, Mead, and Co., 1923), 33–34.

97. Park & Tilford v. Realty Advertising & Supply Co., Court of Appeals, State of New York, 172 App. Div. 955, affirmed. Complaint, 5–6,

98. Sarah Wasserman suggests that such a "descriptive catalog" is related to other encyclopedic literary forms and was one response to encounters with excess. See Mattie Swayne, "Whitman's Catalogue Rhetoric," *Studies in English* 21 (1941): 162–178; and Paul K. Saint-Amour, "Encyclopedic Modernism," in *Tense Future: Modernism, Total War, Encyclopedic Form* (Oxford: Oxford University Press, 2015).

99. Michel Foucault, *The Order of Things: An Archaeology of the Human Sciences* (New York: Random House, 1970), xv.

100. Ibid., xv–xvii.

101. *Park & Tilford*, Appellant's brief, p. 8.

102. *Park & Tilford*, 337.

103. The cycle was understood both by lighting researchers and by advertisers. See, for example, Matthew Luckiesh, *Color and Its Applications* (New York: Van Nostrand, 1915), 144, 279–280; J. M. Shute, *The Lighting of Signs and Billboards*, Lighting Data (series), Bulletin 131 (Harrison, NJ: Edison Lamp Works, November 1921), 13.

104. Foucault, *The Order of Things*, xvii–xviii.

105. Eric Norman Simons, *Successful Retailing* (London: Pitman and Sons, 1926), 17.

106. William Brevda, "How Do I Get to Broadway? Reading Dos Passos's Manhattan Transfer Sign," *Texas Studies in Literature and Language* 38, no. 1 (Spring 1996): 79–114, at 82.

107. "The Newest Thing," 26; Reilly, "The Newest in Electric Signs," 248.

108. Wolfgang Schivelbusch, *The Railway Journey: The Industrialization of Time and Space in the Nineteenth Century* (Oakland: University of California Press, 2014), 60–61. Schivelbusch credited Dolf Sternberger with this insight.

109. These "'simultaneous signs' of the primitive tableau" stood in contrast to the linear narratives, such as chase scenes, that would later come to dominate film. In Noël Burch, *Life to Those Shadows*, trans. Ben Brewster (Berkeley: University of California Press, 1990), 154.

110. On the proto-cinematic aspects of arcades, see Anne Friedman, *Window Shopping: Cinema and the Postmodern* (Berkeley: University of California Press, 1993).

111. Schivelbusch, *The Railway Journey*, 61.

112. Doane drew on Schivelbusch's discussion of panoramic perception: somatic and spatial awareness are subordinated so that film's visual and auditory aspects may dominate. The cinematic facets of Times Square are also discussed in Gordon, *The Urban Spectator*, 79–82.

113. Berman, *On the Town*, 39.

114. Walter Benjamin, "This Space for Rent" (1925–1928), in *One-Way Street and Other Writings*, trans. Edmund Jephcott and Kingsley Shorter (London: NLB, 1979), 89. Anthony Vidler notes that "the collapse of perspective distance" is a dominant theme in Benjamin's writings: Anthony Vidler, *Warped Space: Art, Architecture, and Anxiety in Modern Culture* (Cambridge, MA: MIT Press, 2001), 86.

115. Matthew Luckiesh in 1915 suggested as much when he wrote of the nearly "endless variety of effects" to be gained by combining pictures with flashing letters: Luckiesh, *Color and Its Applications*, 279–280.

116. Wordsworth's list of the types of people he encountered suggests that the listing of signs at Times Square also registered the human traffic of the crowd. My thanks to Sarah Wasserman for clarifying this point. See William Wordsworth, "Residence in London," *The Prelude* (London: Edward Moxon, 1850), 169–204.

117. An Outsider, "The Terrible Advertising Man," *New Republic* 20, no. 250 (Aug. 20, 1919): 85.

118. W. C. Jenkins, "The 'Signs' of the Times," *National Magazine* 34, no. 1 (May 1911): 153–160, at 156.

119. Mildred Adams, "In Their Lights the Cities Are Revealed," *New York Times*, Dec. 11, 1932, SM12, 19.

120. Cited in Leach, *Land of Desire*, 343.

121. Michael Schudson, *Advertising, the Uneasy Persuasion: Its Dubious Impact on American Society* (New York: Basic Books, 1984), 209–233.

122. See, for example, Michael Taussig, "Tactility and Distraction," in *Rereading Cultural Anthropology*, ed. George Marcus (Durham, NC: Duke University Press, 1992), 12; and Marcus Bullock, "Treasures of the Earth and Screen: Todd Haynes's Film 'Velvet Goldmine,'" *Discourse* 24, no. 3 (Fall 2002): 4.

123. Benjamin, "This Space for Rent," 89–90.

124. H. G. Wells, *The Future in America: A Search after Realities* (London: Chapman and Hall, 1906), 58.

125. Edward T. Williams, "Niagara to Illuminate Itself," in *Niagara: Queen of Wonders* (Boston: Chapple Publishing Co., 1916), 160.

126. Edmund Burke, *On the Sublime and the Beautiful*, ed. Charles W. Eliot (New York: P. F. Collier and Son, 1909), 49.

127. E. S. Martin, "As It Rushes By," in W. D. Howells et al., *The Niagara Book* (New York: Doubleday, Page, 1901), 271–73.

128. Robert O. Harland, *The Vice Bondage of a Great City or the Wickedest City in the World* (Chicago: Young People's Civic League, 1912), 67.

129. Wells, *The Future in America*, 77.

130. C. Y. Turner, "The Color Scheme," *Pan-American Art Hand-Book* (Buffalo, NY: David Gray, 1901), 21.

131. Peter A. Porter, *Official Guide Niagara Falls River Frontier* (Buffalo, NY: Matthews-Horthrup Works, 1901), 277.

132. The falls were temporarily lighted several times before, including for the 1901 Exposition. See "Night Illumination of the Falls," in Williams, *Niagara*, 160–165.

133. Gude, "Art and Advertising," 3.

134. Simeon Strunsky, "The Street," *Atlantic Monthly* (February 1914): 221–228, at 228.

135. James C. Young, "Broadway's Own Big Sideshow," *New York Times*, Nov 18, 1928, XX2.

136. Berman, *On the Town*, 9–10. More recently, Cynthia Ozick termed the "white glow that fizzes upward from the city" "an inverted electric Niagara": Cynthia Ozick, *Quarrel & Quandary: Essays* (New York: Vintage International, 2001). Wolfgang Schivelbusch recalled marveling as a child at Times Square, which he knew only from calendars his father received from business associates in America, "including an image of the illuminated artificial waterfall" there. Wolfgang Schivelbusch, "Afterword," in Isenstadt, Neumann, and Petty, eds., *Cities of Light*, 187.

137. Sinclair Lewis, *Main Street* (New York: P. F. Collier and Son, 1920), 416.

138. "New Wrinkles in Electric Signs," *Literary Digest* 71, no. 2 (Oct. 8, 1921): 24.

139. David E. Nye, *American Technological Sublime* (Cambridge, MA: MIT Press, 1994), 179–180.

140. Mortimer P. Reed and E. J. Walcott, "Value of Illumination as an Investment," *Public Service* 9, no. 3 (September 1910): 93–95, at 93.

141. "Value of 'White Way' Street Lighting Improvements," *Electrical Review* 79, no. 2 (July 9, 1921): 58.

142. S. G. Hibben, "The Growth of Ornamental Street Lighting," *Good Lighting* 7, no. 2 (April 1912): 82–84, at 83.

143. Blythe, "The Itch for Publicity," 8.

144. Eugene A. Creed, "Electric Signs in Auburn," *Signs of the Times* 9, no. 38 (June 1909): 16.

145. "Our 'White Way,'" *Los Angeles Times*, Nov. 24, 1910, II4.

146. "A Plan for a White Way," *Dry Goods Reporter* 46, no. 11 (Mar. 20, 1915): 9.

147. H. L. Mencken, *The American Language* (New York: Alfred A. Knopf, 1936), 546.

148. Sagalyn, *Times Square Roulette*, 50–51.

149. Cited in Allen, *The City in Slang*, 59.

150. Chalmers, "How Broadway's Magic Affects the Stranger," SM5.

151. Stephen Graham, *New York Nights* (New York: George H. Doran, 1927), 13.

152. Cited in Still, *Mirror for Gotham*, 260.

CHAPTER 6: GROPING IN THE DARK

An earlier version of this chapter appeared as "Groping in the Dark: The Scotopic Space of Blackouts," *The Senses and Society* 7, no. 3 (November 2012): 302–328.

1. György Kepes, *Light as a Creative Medium* (Cambridge, MA: Harvard University, Carpenter Center for the Visual Arts, 1965), 3.

2. Mable R. Gerken, *Ladies in Pants: A Home Front Diary* (New York: Exposition Press, 1949), 59–61.

3. E. M. Herr, "Accomplishments and Prospects," *Electrical World* 93, no. 1 (Jan. 5, 1929): 21.

4. Wyatt Wells, *Antitrust and the Formation of the Postwar World* (New York: Columbia University Press, 2002), 19–23.

5. A. A. Bright, Jr., and W. R. MacLaurin, "Economic Factors Influencing the Development and Introduction of the Fluorescent Lamp," *Journal of Political Economy* 51 (1943): 339–340.

6. See Harold L. Platt, *The Electric City: Energy and the Growth of the Chicago Area, 1880–1930* (Chicago: University of Chicago Press, 1991), 34–37, 70–71.

7. "The Darkness," *Living Age* 3789 (Feb. 17, 1917): 440.

8. Mildred Adams, "Shadows Cast a Spell Over New York," *New York Times*, Feb. 3, 1929, 79.

9. G. Stanley Hall and Theodate L. Smith, "Reactions to Light and Darkness," *American Journal of Psychology* 14, no. 1 (January 1903): 21–83.

10. William L. Shirer, "Berlin, Oct. 8, 1939," in *Berlin Diary: The Journal of a Foreign Correspondent, 1934–1941* (New York: Alfred A. Knopf, 1942), 233.

11. James M. Landis, "Get Ready to Be Bombed," *American Magazine* 135, no. 6 (June 1943): 30. Landis was director of the U.S. Office of Civilian Defense when he wrote the article. He was remarking on greater ranges of enemy aircraft and the exposure of coastal cities.

12. Even American cities enacted blackouts in World War I. New York, for example, began what came to be called a "calcium curfew," when Broadway's lights were dimmed to such an extent that it "is not Broadway at all. It is Main street somewhere in the middle West." In these instances, though, the primary motivation was to conserve fuel, which could then be directed toward the war effort. "Broadway Puts on Its War Dimmers as Means of Aiding Boys Over There," *Sun*, Nov. 25, 1917, 5.

13. Eugene M. Emme, "Technical Change and Western Military Thought—1914–1945," *Military Affairs* 24, no. 1 (Spring 1960): 10; Ian Patterson, *Guernica and Total War* (Cambridge, MA: Harvard University Press, 2007), 84.

14. Colonel Louis Jackson, "The Defense of Localities against Aerial Attack," *Infantry Journal* 11, no. 3 (November–December 1914): 409–428.

15. "Protection of Civilian Populations against Bombing from the Air in Case of War" (League of Nations, Sept. 30, 1938); Cambridge Scientists' Antiwar Group, *The Protection of the Public from Aerial Attack: Being a Critical Examination of the Recommendations Put Forward by the Air Raid Precautions Department of the Home Office* (London: V. Gollancz, 1937); James T. Muirhead, *Air Attack on Cities: The Broader Aspects of the Problem* (London: George Allen and Unwin, 1938); C. W. Glover, *Civil Defense: A Practical Manual Presenting with Working Drawings the Methods Required for Adequate Protection against Aerial Attack* (London: Chapman and Hall, 1938).

16. Paul K. Saint-Amour, *Tense Future: Modernism, Total War, Encyclopedic Form* (Oxford: Oxford University Press, 2015), 34–35.

17. Robert A. Divine, *Roosevelt and World War II* (Baltimore, MD: Johns Hopkins University Press, 1969), 8.

18. Leo Mellor, "Cityscape: The Bombed City in the Second World War," in *The Edinburgh Companion to Twentieth-Century British and American War Literature*, ed. Adam Piette and Mark Rawlinson (Edinburgh: Edinburgh University Press, 2012), 456. See also Ian Patterson, *Guernica and Total War* (London: Profile, 2007); or, from a more military-historical standpoint, Tami Davis Biddle, *Rhetoric and Reality in Air Warfare: The Evolution of British and American Ideas about Strategic Bombing, 1914–1945* (Princeton, NJ: Princeton University Press, 2002); John H. Morrow, Jr., "States and Strategic Airpower: Continuity and Change, 1906–1939," in *The Influence of Airpower upon History: Statesmanship, Diplomacy, and Foreign Policy since 1903* (Lexington, KY: University Press of Kentucky, 2013), 37–60; Ward Thomas, "Aerial Bombing to 1945: 'A Frightful Cataclysm,'" in *The Ethics of Destruction: Norms and Force in International Relations* (Ithaca, NY: Cornell University Press, 2001), 87–146; and Christopher Joel Simer, "Apocalyptic Visions: Fear of Aerial Attack in Britain, 1920–1938" (Ph.D. diss., University of Minnesota, 1999).

19. The phrase "medieval flavor" is from Sir Philip Gibbs, "War Rigors Alter Habits of Britain" *New York Times*, Sept. 27, 1939, 9; "cave-man's existence" is from "Mental Health in Wartime," *British Medical Journal* 1, no. 4179 (Feb. 8, 1941): 208. See also "Gibraltar to Be Darkened Tonight," *New York Times*, Oct. 3, 1935, 3; "Istanbul Douses Lights in First Air Raid Drill," *New York Times*, Dec. 21, 1935, 8; "Reich Gets Ready for 'Black Week,'" *New York Times*, Sept. 19, 1937, 21; Hugh Byas, "Tokyo to Combat 'Air Raids' Today," *New York Times*, Sept. 15, 1937, 13; and P. J. Philip,

"Capitals of Warring Countries Struggle to Be Gay on the First Blacked-Out Christmas in History" *New York Times*, Dec. 25, 1939, 10.

20. Winston Churchill, "'The Few,' August 20, 1940," in *Winston S. Churchill: His Complete Speeches: 1897–1963*, ed. Robert Rhodes James, vol. 6, *1935–1942* (New York: Chelsea House, 1974), 6266.

21. See discussion in E. C. Daniel, "How Would America Defend Itself against Air Attack?," *Washington Post*, June 2, 1940, B4; and "Is Your Water-Works Plant Prepared?," *American City*, August 1942, 9.

22. "Northeast to See 'Warlike' Army Air Show," *New York Times*, Apr. 25, 1938, 1; "Town to 'Black Out' in 'Air Raid' Tonight," *New York Times*, May 16, 1938, 1; "Blackout in Hawaii Foils 'Enemy' Fliers," *New York Times*, May 20, 1939, 7. For an overview of American blackouts at this time, see David E. Nye, *When the Lights Went Out: A History of Blackouts in America* (Cambridge, MA: MIT Press, 2010), 37–66.

23. James Reston, "All of Life in Britain Transformed by War," *New York Times*, Sept. 10, 1939, E5.

24. Edward R. Murrow, "This … Is London," CBS radio broadcast, Aug. 24, 1940.

25. Brooks Atkinson, "Going to the Fair: Spectacle of People in Carnival Mood Has New Meaning This Year," *New York Times*, May 19, 1940, 131.

26. Matthew Dallek, *Defenseless under the Night: The Roosevelt Years and the Origins of Homeland Security* (New York: Oxford University Press, 2016), 115–128.

27. "Women Now Face Grim Task Calmly," *New York Times*, Dec. 20, 1941, 23.

28. For example, see Roger W. Lotchin, "California Cities and the Hurricane of Change: World War II in the San Francisco, Los Angeles, and San Diego Metropolitan Areas," *Pacific Historical Review* 63, no. 3 (August 1994): 393–420.

29. "FDR and the Media," *U.S. Presidents as Orators: A Bio-Critical Sourcebook*, ed. Halford Ryan (Westport, CT: Greenwood Press, 1995), 157–159.

30. Shirer, *Berlin Diary: The Journal of a Foreign Correspondent, 1934–1941*, 198.

31. "Great Signs Dark as Gay White Way Obeys Army Edict," *New York Times*, Apr. 30, 1942, 1. See also "Blackout for the White Way," *Rockefeller Center Magazine*, September 1942, 17.

32. Walter Miles, "Night Vision," *Yale Scientific Magazine*, October 1943, 10.

33. "How to See at Night," *Life*, Oct. 5, 1942, 71.

34. S.G. Hibben and K. M. Reid offer a survey of visual perception studies during blackouts in "Comments on Blackouts," *Illuminating Engineering* 37 (1942): 210–216.

35. Oskar Nagel, "On Seeing in the Dark," *Psychological Review* 15 (July 1908): 250–254.

36. Blackout Hard on City Eyes, *Washington Post*, Apr. 26, 1942, L9. See also Selig Hecht, "Seeing in a Blackout," *Harper's*, July 1942, 160.

37. Curt Wachtel, *Air Raid Defense (Civilian)* (Brooklyn, NY: Chemical Publishing Co., 1941), 76–77.

38. Géza Révész, "The Problem of Space with Particular Emphasis on Specific Sensory Spaces," *American Journal of Psychology* 50 (1937): 434–4.

39. Walter Blumenfeld, "The Relationship between the Optical and Haptic Construction of Space," *Acta Psychologica* 2 (1937): 125–174.

40. "First Aid in Air Raids: Experiences of Bristol Practitioners," *British Medical Journal* 1, no. 4199 (June 28, 1941): 978–979.

41. James Hopper, "Blackout," *Los Angeles Times*, Sept, 5, 1943, G8.

42. These and similar reports recall an earlier use of the phrase "common sense," a postulated mental faculty that integrated sensory impressions into a coherent perception of the world. Following the work of nineteenth-century physiological psychologists, the singular faculty of "common sense" had been eclipsed by an ongoing process of apperception, the mental process of assimilating new experiences in terms of prior ones in order to make sense of them, as a "relating activity" among distinct senses, as the philosopher and psychologist John Dewey had once put it. John Dewey, *Psychology* (New York: Harper and Bros., 1886), 89, and "Attention," 132–147.

43. Richard Shusterman, "Somaesthetics: A Disciplinary Proposal," *Journal of Aesthetics and Art Criticism* 57, no. 3 (Summer 1999): 299–313.

44. Blumenfeld, "The Relationship between the Optical and Haptic Construction of Space," 130.

45. Hecht, "Seeing in a Blackout," 160.

46. "Second Blackout Darkens Brooklyn," *New York Times*, Apr. 22, 1942, 14.

47. Peter C. Baldwin, "In the Heart of Darkness: Blackouts and the Social Geography of Lighting in the Gaslight Era," *Journal of Urban History* 30, no. 5 (July 2004): 749–768.

48. Karl F. Muenzinger, *Psychology: The Science of Behavior* (New York: Harper and Bros., 1942).

49. Blumenfeld, "The Relationship between the Optical and Haptic Construction of Space,"134. The tactile nature of night has long been noted in other contexts, too. See Christopher Dewdney's discussion in *Acquainted with the Night: Excursions through the World after Dark* (Toronto: HarperCollins, 2004), 137.

50. *December 7*, John Ford, director (U.S. War Department and U.S. Navy Department, ca. 1943).

51. Edwin Boring, "Panic and Mobs," *Psychology for the Armed Service* (Washington, DC: Infantry Journal, Committee of the National Research Council on a Textbook of Military Psychology, 1945), 445.

52. Kirke Simpson, "Blackout Toll May Prompt Active War," *Washington Post*, Jan. 21, 1940, 11. Wilfred Trotter estimated 600 fatal accidents might be attributed to blackouts and the panic associated with them: Wilfred Trotter, "Panic and Its Consequences," *British Medical Journal*, Feb. 17, 1940, 270. The definitive study of traffic accidents in Britain's blackouts is H. M. Vernon, *Road Accidents in Wartime* (Oxford: Heffer and Sons, 1941).

53. "Blackout Raising New Psychological Problems," *Science News-Letter* 38, no. 4 (July 27, 1940): 57; "How to Guard against Home Blackout Accidents," *Science News-Letter* 41, no. 6 (Feb. 7, 1942): 84–85.

54. Franz Pick, "Black-Outs Come to America," *Barron's*, Dec. 15, 1941, 21, 50, and May 26, 1941.

55. Britain's very different circumstances did apparently lead to more crime, notably assaults, burglaries, and short-changing of customers. Victor H. Evjen, "Delinquency and Crime in Wartime," *Journal of Criminal Law and Criminology* 33, no. 3 (July–August 1942): 140.

56. Lowell S. Selling, "Specific War Crimes," *Journal of Criminal Law and Criminology* 34, no. 5 (January–February 1944): 303–310.

57. Quoted in James Hopper, "Blackout," *Los Angeles Times*, Sept. 5, 1943, G8.

58. V. J. Batteson, "'Black-out' Problems," *British Medical Journal*, Oct. 21, 1939, 831. To emphasize his point, the writer cited the melancholic Milton of *Il Penseroso*: "Where glowing embers through the room / teach light to counterfeit a gloom." See also M. Forrest, "Lighting in the Home: Counteracting Blackout Depression," *Electrical Review* 127 (Sept. 6, 1940): 197ff.

59. C. E. Ferree and Gertrude Rand, "The Eye as a Factor in Wartime and Blackout Lighting," *Journal of General Psychology* 29 (1943): 281–301.

60. R. Lindsay Rea, "Night Blindness: A Further Warning," *British Medical Journal* 1, no. 4129 (Feb. 24, 1940): 319; and F. M. R. Walshe, "Night-Blindness: A Psychological Study," *British Medical Journal* 2, no.4223 (Dec. 13, 1941): 858.

61. Brett Holman, "The Air Panic of 1935: British Press Opinion between Disarmament and Rearmament," *Journal of Contemporary History* 46, no. 2 (April 2011): 288–307.

62. "Blackout Raising New Psychological Problems," 57. Summarized from a study by K. J. W. Craik, published in the *Scientific Worker* in June 1940. Craik continued to study the effect of affective states on visual perception in low light conditions: Craik and M. D. Vernon, "Perception during Dark Adaptation," *British Journal of Psychology* 32 (January 1942): 206–230. A third trigger, according to some British sources, was racial and cultural. Jews, for example, were said by some at the time to be especially likely to panic. Geoffrey Field, "Nights Underground in Darkest London: The Blitz, 1940–1941," *International Labor and Working-Class History* 62 (Fall 2002): 14, 19.

63. Boring, "Panic and Mobs," 462. Waiting is in many ways the subject of Archibald MacLeish's 1938 radio drama, "Air Raid."

64. Ibid., 456–457. See also Hadley Cantril, *The Invasion from Mars: A Study in the Psychology of Panic* (Princeton, NJ: Princeton University Press, 1940).

65. Wilfred Trotter, "Panic and Its Consequences," *British Medical Journal*, Feb. 17, 1940, 270.

66. Elizabeth Bohm, "Lighted Window," *Saturday Evening Post* 216 (Jan. 8, 1944): 53.

67. Ethel Lina White, "Blackout," *Chicago Tribune*, Aug. 25, 1940, 99.

68. Norman Rockwell, *What to Do in a Blackout*, *Saturday Evening Post*, June 27, 1942, cover.

69. "Taste for the Dark," *Washington Post*, Oct. 31, 1940, 10.

70. Boring, "Panic and Mobs," 455–456, 462–463. Commonly referenced studies of the time include Gordon W. Allport and Helene R. Veltfort, "Social Psychology and the Civilian War Effort," *Journal of Social Psychology* 18 (1943): 165–233; and P. E. Vernon, Psychological Effects of Air-Raids," *Journal of Abnormal and Social Psychology* 36 (1941): 457–476.

71. Gordon W. Allport and Gertrude Schmeidler, "Morale Research and Its Clearing," *Psychological Bulletin* 40 (1943): 65–68, at 68.

72. Merl Bonney, "Psychological Factors in Total Fitness for War," *Journal of Health and Physical Education* 14, no. 5 (May 1943): 255. For more on psychological research into civilian response to wartime, see Dietmar Süss, "The Struggle to Win Trust and Maintain Morale" and "Wartime Morale as an Object of Research" in *Death from the Skies: How the British and the Germans Survived Bombing in World War II*, trans. Lesley Sharpe and Jeremy Noakes (Oxford: Oxford University Press, 2014), 45–59, 353–366.

73. "New Blackout Methods," *Architect and Engineer* 152, no. 3 (March 1943): 43.

74. See Jerry A. Woolley, *Point Pleasant* (Charleston, SC: Arcadia Publishing, 2003), 119.

75. Ray Shaw, "Fun in a Blackout." *American Magazine* 135 (March 1943): 52–53.

76. "Blackout Games," *Woman's Home Companion* 69 (April 1942): 104; Grace Pennock, "It's Not the Cold but the Dark," *Ladies' Home Journal* 59, no. 21 (February 1942): 102; "Life Goes On behind the Blackout," *Popular Mechanics*, January 1943, 88–89.

77. See the issue of *School Arts* 43, no. 2 (October 1943). For more on cheering up blacked-out homes, see Walter Rendell Storey, "Trends in Home Decoration," *New York Times*, Aug. 23, 1942, D4.

78. W. H. Auden, "September 1, 1939," in *Selected Poems*, ed. Edward Mendelsohn (New York: Vintage, 1979), 89. Originally published in *The New Republic*, October 18, 1939.

79. "No Blackout," *Christian Science Monitor*, May 27, 1940, 3.

80. Ward Harrison, "Blackouts," *Transactions of the Illuminating Engineering Society* 38, no. 4 (April 1943): 165–166.

81. Radio address delivered by President Roosevelt from Washington, DC, Sept. 3, 1939.

82. President Franklin Roosevelt's "The Light of Democracy Must Be Kept Burning" speech, March 15, 1941; *The Public Papers and Addresses of Franklin D. Roosevelt*, 1941 vol., *The Call to Battle Stations,* 60–69.

83. *Chicago Tribune*, Dec. 29, 1940, 7.

84. Discussed in Richard Snow, *A Measureless Peril: America in the Fight for the Atlantic* (New York: Scribner, 2010), 154–155.

85. General Electric advertisement, *American Magazine* 135 (1943), n.p.

Index

Note: Page numbers in *italics* refer to illustrations.

A&P's Eight O'Clock brand coffee, 161
Accidents
  during blackouts, 230–231
  with electricity, 32
Acetylene lamps, 7, 68, 72, 83
Action
  at a distance
    of electric lighting, 14–15, 25, 26
    of switches, 49
  modern desire for, 181
Adams, Henry, 60–61
Adams, Henry Foster, 180
Adams, Mildred, 194–195, 209, 211
Adorno, Theodore, 169, 178
Advertisements. *See also* Advertising;
    Commercial speech
  for A&P's Eight O'Clock brand coffee, 161
  of Benjamin Industrial Lighting, *139*
  for Bond clothing store, 199, *199*
  for Camel cigarettes, 169
  of Corning Glass Works, 84
  of Corticelli Silk Mills, 167
  of General Electric Co., 13, *33*, *62*, *68*, *100*, *104*, *128*, *136*, *150*, *212*, *239*, *240*
  of Habirshaw Company, *57*
  for Heatherbloom petticoats, 167, *168*
  of Heinz Company, *162*, 163
  for Legalite Lenses, Hyslop Bros., *97*
  of Lightolier Co., *54*
  for Mazda automobile headlamps, *68*
  for Mazda Lamps, *128*, *136*, *212*
  of National Electric Light Association, *48*
  of *New York Times*, 163
  of O. J. Gude Company, *160*
  of Osgood lenses, 99, *99*
  of Prest-o-lite Co., *74*
  of Rice Electrical Company, 169, *173*
  of Rumely Products Co., *36*
  for Rural Electrification Administration, *210*
  for Second National Swiss Automobile Exposition (Geneva), *90*, *91*
  for Tally-Ho Bicycle Lanterns, *69*
  for Tropicana products, 169
  for Westinghouse Individual Motor Drive, *124*
  of Westinghouse Lamps Co., *134*
  for White Rock Table Water, 167, 169
  for Wrigley's Spearmint Chewing Gum, 169, *170–171*
Advertising. *See also* Advertisements;
    Commercial speech
  attention/memory value in, 180, *180*
  criticism and, 196
  motion theory in, 179
  psychology of, 180
  repetition in, 179, *180*
Aerial bombing, 207, 213, *214*, 217, 233, 238
Agency, switches and, 35, 60, 61
American Architect, 132
American business, 194
American character, 194–195, 198
American Institute of Electrical Engineers, 7
American Journal of Psychology, 211
American Optometric Association, 147
American Posture League, 149
Amoskeag Manufacturing Company, 123, 132
"The Analytical Language of John Wilkins" (Borges), 189
Animation
  of commercial speech
    pictures, 167, *168*, 169, *170–171*, *172*, *173*, *175–176*
    texts, 163, *164*, *165*, *166*, 167
Antiglare devices, 96
Apprehension, visual, 126–127, *127*, *128*, 129
Architects
  electric lighting and, 2, 7, 9, 19
  illuminating engineers and, 8, 9
Architecture
  daylight and, 12
  electric lighting as, 1–2
  glare and, 17
  hierarchy in, 194

Architecture (cont.)
  illuminating engineers and, 9
  modernism in, 2, 11–12, 51
  motographs and, 165, *166*, 167
Arc lamps, *4*, *5*, 68, 120–121, *122*
Arkwright, Richard, 118
Armstrong, William, 5
Artificial lighting. *See also specific forms*
  developments in, 2, *3*
  for nighttime labor, 112
Artkraft Strauss, 161
Asia, 203, *204*, 213, 215
Assured clear distance ahead, 80
Asthenopia, 92
Atkinson, Edward, 142
Atlanta Constitution, 100
Attention
  definitions of, 129–130
  efficiency of, 130, *130*, 132
  value, 178–181, *180*
Auburn (NY), 201
Auden, W. H., 215, 237
Augustine, Reginald, 147
Automobile Club (Massachusetts), 94
Automobiles
  batteries of, 99
  braking distance of, *79*, 80–81, *82*, 83
  drivers of (*see* Drivers)
  driving of (*see* Driving)
  as ferocious animals, 87
  gender of, 89
  growth in number of, 72
  lamps on, 72 (*see also* Headlamps/ headlights)
  regulations on, 81

Back of the Button (film), 59
Baines, Edward, 118, 120
Bakers, 109
Baldwin, Peter, 229
Baldwin, Stanley, 213, 232
Baldwin, William, 109
Balla, Giacomo, *10*, 11
Barbour, Bill, 231
Barron's, 230–231
Batteries, of automobiles, 99
Baudelaire, Charles, 152
Beardsley, Aubrey, *69*

Beaver Falls (PA), 43
Belt-and-shaft systems, 113, 124–125
Benjamin, Walter, 192, 193, 196
Benjamin Industrial Lighting, *139*
Bennette, Arnold, 186
Bentham, Jeremy, 152
Berlin: Symphony of a Great City (film), *202*
Berman, Marshall, 155, 193, 199
Bernays, Edward, 209
Bernhard, F. H., 137
Bête humaine, La (Zola), 68
Better Light–Better Sight Bureau Press, 149
Better Light–Better Sight campaign, 149
Better Light–Better Sight Lamps, 149
Bickley, Everett, 165
Bicycle lamps, 68, *69*
Bicycles, 89
Blackout (game), 237, *238*
Blackout depression, 231
Blackouts (wartime). *See also* Darkness
  accidents during, 230–231
  in Asia, 213, 215
  criminality during, 231
  erotic side of, 233, 235–236, *235*
  in Europe, 213, 215
  in general, 227, 229
  in Great Britain, 215, 217, 230–232
  homes/home life and, *228*, 229, 237
  maintaining routines and, 236–237
  mystery stories and, 235
  pathological conditions and, 231–233
  regulations on, violations of, 229, 231
  rehearsals of, 239
  in United States, *71*, 207, *208*, 215, 216, 217, *218–219*, 224, 231–232, *234*, 236, 238–240
Blindness
  night, 92, 232
  regaining sight after, 230
  temporary, 67, 87
Blumenfeld, Walter, 223, 225, 230
Boccioni, Umberto, 1
Bohm, Elizabeth, 233
Bohm, Max, 233
Bohrer, Karl Heinz, 61
Bolton, Charles Edward, 40
Bolton, Sarah Knowles, 40
Bond clothing store, 199, *199*

Boott Cotton Mills (Lowell, MA), *115*
Borges, Jorge Luis, 189, 190
Borgmann, Albert, 47, 60
Boring, Edwin, 230, 232–233, 236
Boulder Dam, 43, 45
Bradley, Milton, 237
Braking distance
 agreement on safe, 81
 of automobiles, *79*, 80–81, *82*, 83
 definition of, 81
 of locomotives, 79–80
 at night, 83
Brandeis, Louis, 33
Brazil, 25
Breasts, headlamps and, *90*, 91
Brevda, William, 191
British Medical Journal, 140
British Ministry of Information, 236
Broadway (New York). *See also* Times Square (New York)
 electric lighting of, and commercial speech
  attention/memory value and, 180–181
  as cinema, 177
  comparisons to, 203
  in general, 22, 155, 157, *158*
  picture animation of, 169, *173*
  street lights on, 11, 159
  text animation of, 163
 gas lighting of
  commercial speech on, 157, 159
  street lights on, 157
Brooke, Rupert, 186–187, 193
Brooklyn Woman's Club, 53
Brush, Charles F., 120–121
Buffalo (NY), 198
Bukit Bintang (Kuala Lumpur), 203
Bulbs. *See* Lamps/lights
Burch, Noël, 192
Bureau of Education, 149
Burke, Edmund, 196–197
Bursiel, Miss, 59

Calculable man, 132
California Automobile Trade Association, 102
Camel cigarettes, 169
Cameras, 40
Candles, 2, 112, 118

Carriage lamps, 68
Cars. *See* Automobiles
Catholic Church, 25
Chalmers, Stephen, 176, 203
Chariot race sign, 169, *173*, 177, 190
Chesterton, G. K., 187, 191
Chicago
 entertainment district in, 197
 Knights Templar sign in, *172*
 World's Fair in, 37, *38*, 39–40, 163
Chicago Tribune, 233, 235
China, 215
Christian Science Monitor, 238
Christmas trees, 42, 239, *240*
Cinema
 Broadway as, 177
 streets as, 176, 177–178
 Times Square as, 192–193
Cities/city life. *See also under specific cities*
 commercial speech in, 157
 driving in, 77–78
Classes, headlight adjusting, 103
Claude, Georges, 161
Clayton, John, 109
Cleveland, Grover, 37, *38*, 39–40
Clewell, Clarence Edward, 134, 135, 145
Cline, Harry X., 72
Cobblers, 109
Code of Lighting for Factories, Mills and Other Work Places, 137
Cologne, 213
Colorado River, 43
Commercial speech. *See also* Advertisements; Advertising
 in city life, 157
 electric lighting of
  animation of pictures, 167, *168*, 169, *170–171*, *172*, *173*, 175–176
  animation of texts, 163, *164*, 165, *166*, 167
  attention/memory value and, 180–181
  compulsory aspect of, 185
  in New York (*see* Broadway [New York]; Times Square [New York])
  in Paris/London, 157 (*see also* Great White Ways)
 gas lighting of, 157, 159

INDEX 281

Commonwealth Edison (Chicago), 135
Concentration, night driving and, 75–77, *76*, 103
Congressional Library, *16*
Coolidge, Calvin, 42
Corbin, Austin, 163
Corning Glass Works, 84
Corticelli Silk Mills, 167
Cost, of lighting, 132
Costing, live-cycle, 132
Country roads, 77–78, *78*
Countryside, darkness of, 77
Court cases, on glare, 94
Crane, Stephen, 159
Crane, Walter, *36*
Crary, Jonathan, 77, 129
Criminality, 231
Cristo Redentor, 25, *26*
Criticism, advertising and, 196
Cromford Mill (Derbyshire), 118, *119*
Crooks, Laurence E., 67, 103, 105
Crystal Palace (London), 163
Cubberley, Ellwood P., 146
Cubism, 155
Cumberland Hotel, 163
Cyclorama, 192

Darkness (wartime). *See also* Blackouts
  between flashes, 189–190
  headlights and, 75, 77, 101
  nostalgia for, 209, 211
Dark Victory (film), 240
Davis, Bette, 240
Davis, Lavinia Riker, 101
Daylight, 12, 113–114, *116*, *117*
Daytime labor, 109
December 7 (film), 230
Dekker, Thomas, 112
De Segundo, Edward, 27
Designs
  of industrial facilities, *115*, *124*, *125*
  of light switches, 28, *29*, 31, *31*
Detroit (MI), 179
Device paradigm, 46–47, 49, 60
Dewey, John, 129–130
Diana (St. Gaudens), 159
Distance, action at a, 14–15, 25, 26, 49
Doane, Mary Ann, 193

Dow, J. S., 129
Dreiser, Theodore, 163
Drivers
  of automobiles
    as ferocious animals, 87–88
    legal responsibilities of, 80–81
    managing of repairs by, 102
    reaction time and, 80
    technical skills of, 102–103
Driving
  in cities, 77–78
  in countryside, 77–78, *78*
  experiencing space while, 105
  at night
    concentration and, 75–77, *76*, 103
    as entertainment, 75, 101–102
    glare and (*see* Glare)
    joyriding and, 89
    relaxing effect of, 77
    speed and, 83, 87–88, 103
    visual aspects of, 73, *73*, 74, 75
    by women, 92
  by women, 91–92, *104*
Dry Goods Reporter, 178
Dull, Egbert Reynolds, 163, 165
Durgin, William, 132, 135
Düsseldorf, 213
Dynamos, 60–61

Eastman Kodak, 40
Edison, Thomas
  in general, 5, 13, 14
  patents of, 28, *29*, *30*, 31
Edison Company, 121
Edison Electrical Institute, 149
Edison Monthly, 157
Edison Tower, 40, *41*
Edwards, Evan, 83
Efficiency
  of attention, 130, *130*, 132
  of laborers, *141*
  speed and, 126–127
Election results, *175*, 176
Electrical industry, standardization of, 136–137, *136*
Electrical Review, 113, 126
Electrical switches. *See* Switches/buttons
Electric House, The (Keaton), 51–52, *52*

Electricity. *See also* Electric lighting; Electrification
  access to, 5, 7
  accidents with, 32
  commercial distribution of, 5
  explanations of, 32
  generation of, 43
  God and, 25, 26, 39–40
  invisibility of, 25, 46–47
  magnetism and, 14–15
  national cohesion and, 45
  women and
    explaining to, 56, 59
    incompetence with, 56
Electricity grid, 5, 45
Electric Journal, 132
Electric lighting. *See also* Lamps/lights
  action at a distance of, 14–15, 25, 26
  American perception of, 239–241
  architects and, 2, 7, 9, 19
  as architecture, 1–2
  characteristics of, 14, 25
  of Christmas trees, 42, 239, *240*
  commercial speech and (*see* Commercial speech)
  cost of, 132
  of *Cristo Redentor*, 25, *26*
  as entertainment (*see* Entertainment)
  geometry of, *130*
  glare of (*see* Glare)
  history of, 2–3, *3*, *4*, 5, 7
  of homes (*see* Homes/home life)
  of industrial facilities (*see* Industrial facilities)
  instantaneity of, 14, 15, 25, 26
  intensities of, 127, *127*, 129, 142, 151, 209, 232
  loss of (*see* Blackouts [wartime])
  material progress and, 7, 161, 209
  measurement of, 149–151
  modernism and, 9, 11–12
  of Niagara Falls, 198
  of offices, 151
  productivity and, 125, 132–133, *134*, 135, 149
  recoding of, 229
  of rooms, 59–60
  of schools, 146–147, *148*, 149
  in sequence, 35
  of shops, 146
  of streets (*see* Streets)
  switching on (*see also* Switches/buttons)
    as entertainment, 37, *38*, 39–40, *41*
    in political performances, 42, *42*, 43
    representational powers of, 239
  uplifting effect of, 211
  value of, 132–133, *134*, 135
  visual apprehension and, 126–127, *127*, *128*, 129
Electric Light's Golden Jubilee, 209
Electric modernism, 9, 11–12
Electric motors, 124–125, *124*
Electrification, 2–3, 5, 7, 43
Elliott, Elias Leavenworth, 9, 125, 137, 161
Engels, Friedrich, 140
Entertainment
  driving by night as, 75, 101–102
  electric lighting as, 37, *38*, 39–40, *41*, 176, 177–178, 192–193
Epiphany, 61
Epok system, 169, 175, 177
Erotic side, of blackouts, 233, 235–236, *235*
Evans, Walker, *184*
Exposition Universelle (Paris), 60–61
Eyes. *See also* Vision/seeing
  evolution of, 129, 146–147, *147*
  of laborers, *138*, 140
  nighttime vision of, *220*, 221–222

Factories. *See also* Industrial facilities; Textile mills
  emergence of, 112–113
  schools as, 107, 109, 146
Factory Lighting (Clewell), 135
Fairbairn, William, 113
Faraday, Michael, 14–15
Farmers, 109
Fatigue
  in general, 140–141
  visual, 141–142
Fear, 233
Federal Aid Road Act (1916), 94
Ferree, Clarence, 232
Field theory, 15
Film, switches in, *52*
Fitch, James Marsden, 126

Flashing signs
 advertising value of, 178
 in general, 163, 165, 167, 169, *172*, *173*, 193
 on-off cycle of, 189–190
Fluorescent light, 209
Foote, Allen Ripley, 121
Ford, Model T, 72
Ford Factory (Highland Park, MI), 125
Fort-da game, 34
Foucault, Michel, 20, 130, 132, 152, 189, 190
France. *See* Paris
Freud, Sigmund, 34, 63–64
Funke, Jaromir, 185

Gaetjens, Allen Kiefer, 112
Gardiner, Sarah Diodati, 67
Gas lamps
 acetylene, 7, 68, 72, 83
 in general, 28, 118
Gas lighting. *See also* Gas lamps
 of commercial speech, 157, 159
 cost of, 132
 spread of, 7
 of textile mills, 118, *119*, 120, 121
Gaster, Leon, 146
Gastineau, Benjamin, 192
General Electric Co.
 advertisements of, *13*, *33*, *62*, *68*, *100*, *104*, *128*, *136*, *150*, *212*, *239*, *240*
 Electric Light's Golden Jubilee, 209
 headlamp made by, 98, 99
 light conditioning campaign of, 150–151
General Services Administration, 151
Gerken, Mable R., 207
Germany, 213, 215
Gibson, James Jerome, 67, 81, 103, 105
Giedion, Sigfried, 105
Gilbreth, Frank, 129, 141–142
Gilbreth, Mary, 129
Glare
 architecture and, 17
 of electric lighting
  in general, *16*, 17, 18, 92, 108
  as ideological problem, 19
  technical progress and, 20
 of headlamps
  antiglare devices, *96*, 98, *98*
  court cases on, 94
 definition of, 94
 in general, *82*, 83–84, *84*, *85*, *86*, 92, *95*
 visual conditions leading to, 87
 women and, 89, 91
 meanings of, 17–18
 sensitivity to, 97
Glasses, glare minimizing, *98*
God, electricity and, 25, 26, 39–40
Goddard, C. M., 120
Goggles, glare minimizing, *98*
Goldberger, Paul, 162
Good Roads movement, 72
Graham, Stephen, 203
Gravity, 14
Great Britain, 215, 217, 230–232. *See also* London
Great Depression, 7, 161, 209
Great White Ways
 in Kabukicho (Tokyo), *204*
 in United States
  Auburn, 201
  in general, 200–201, 203
  Los Angeles, 201
  New Haven, *201*
  New York City, *160*, 161, 177, *195*, 197, 199
 worldwide, 203–204
Gude, Oscar J., 159, 161, *162*, 163, 177, 185, 198–199

Habirshaw Company, *57*
Hague, The, 213
Hall-Brown, Lucy, 53, 55
Hammer, William Joseph, 163
Harrison, Ward, 126, 238
Hay, Alfred, 2
Headlamps/headlights
 of automobiles
  acetylene, 7, 68, 72, 83
  adjusting of, 102–103
  advertisements for, *104*
  breaking distance and, 83
  electric, 72, 83
  in general, 72, 76, 88
  glare elimination built into, 98–99
  glare of (*see* Glare)
  municipal regulations on, 83, 92–93
  repairs to, *104*

sealed beam, 100, *100*
self-propulsion and, 105
shading of, *224*
space of, *81*, *82*, *85*, 102, 103, *104*
standardization of regulations, 97, 100
variations in regulations, 92–93
women's breasts and, *90*, 91
darkness and, 75, 77, 101
in general, 67, 68
of locomotives, 68, *69*, 70
Headlights. *See* Headlamps/headlights
Heaney, Seamus, 25
Heatherbloom petticoats, 167, *168*
Heidegger, Martin, 60
Heinz Company, *162*, 163
Helmholtz, Hermann von, 140
Henderson, Mary, 157
Hero of Alexandria, 27
Herr, E. M., 209
Heterotopias, 190
Highways, lighting of, 100
High, Wide, and Handsome (musical), 175
History, of electric lighting, 2–3, *3*, *4*, 5, 7
Home economists, 56, 59
Homes/home life
  blackouts and, *228*, *229*, *237*
  electric lighting of
    in general, 19, 35, 109, 146
    promotional campaigns for, 150–151, *150*
    switches and, 49, *50*, 51–53
Hoover, Herbert, 42
Horkheimer, Max, 169, 178
Housekeepers, 55
Housework, 51, 53
Hughes, Rupert, 159, 186

Ignorance, switches and, 60
Illuminating Engineer, 161
Illuminating Engineering Society (IES), 149, 209. *See also* International Commission on Illumination (of IES)
Illuminating engineers
  architects and, *8*, 9
  architecture and, 9
  glare and, 17, 100
  measuring light by, 114
  productive vision and, 107, 125, 132

Illuminating Glassware Guild, 149
Incandescent light, 5, *6*, 7, 121, 163, 200, 209
Independent, The, 37
Industrial facilities
  belt-and-shaft systems in, 113, 124–125
  design of, *115*, 124, 125
  electric lighting of
    electric motors and, 125
    in general, 107, *108*
    productivity and, 132–133, *134*, 135
    schemes for, 107, 125–126, 132, 143, *143*, *144*, 145, 232
    standardization of, 137
    stimulating effect of, 142
    as subjective experience, *131*
    textile mills, 120–121, *122*, 123–124
    visual fatigue and, 141–142
    worker safety and, 140
  electric motors in, 124–125, *124*
Industrial Fatigue Research Board, 141
Infinity, symbol of, 61
Instantaneity
  of electric lighting, 14, 15, 25, 26
  of switches, 49, 60, 61, *62*, 63
  of telegraphy, 15
Insull, Samuel, 5
Insurance companies, 118, 121, 132, 137
Intensities
  of electric lighting, 127, *127*, 129, 142, 151, 209, 232
  emotional, 61
  productive, 135
International Commission on Illumination (of IES), 6, 7, 9, 92, 97, 135, 137
International Photometric Commission, 7
Invisibility, of electric lighting, 25
Iron Trade Review, 123
Isherwood, Christopher, 215
Isolationists, 239
Iwo Jima, 177

James, Winifred Lewellin, 14
Japan, *204*, 213, 215
Johnson, Edith, 92
Joyce, James, 61
Joyriding, 89

Kabukicho (Tokyo), *204*
Kearney (NE), 238
Keaton, Buster, 51–52, *52*
Keenan, W. J., 32
Keene (NH), *216*
Kelly, Richard, 25
Kennedy, John F., *45*
Kennelly, A. E., 49
Kepes, György, 207
Kerosene, 68, 132
Kerwer, Fred, 175
Keun, Odette, 167, 181
Knights of Pythias convention, 46
Knights Templar sign, *172*
Kracauer, Siegfried, 152
Kuala Lumpur, 203

Labor
  daytime, 109
  division of, 113
  nighttime
    artificial lighting for, 112
    in general, 109, *111*
    of women, 109, *110*, 112
  visual, 107
Laborers
  efficiency of, *141*
  eyes of, *138*, 140
  fatigue of, 140–141
  safety of, 140
  supervision of, 145
Labor laws, 137
Ladies' Paradise, The (Zola), 18–19
Lamps/lights. *See also* Headlamps/
    headlights
  arc, 4, 5, 68, 120–121, *122*
  on automobiles, 72
  on bicycles, 68, *69*
  on carriages, 68
  fluorescent, 209
  gas
    acetylene, 7, 68, 72, 83
    in general, 28, 118
  glare of (*see* Glare)
  incandescent, 5, 6, 7, 121, 163, 200, 209
  light emitted by, 14
  on locomotives, 68, *69*, 70
  oil, 27, 28, 68, 112, 118

Lancaster System, 146
Landis, James M., 213
Lane-Fox, St. George, 27
Lang, Fritz, 185, *188*
Las Vegas (NV), 203
"Leaders of the World" sign, 169, *173*
League of Nations, 213
Lefebvre, Henri, 79
Leff, Arthur, 80
Legalite Lenses, Hyslop Bros., 97
Leigh, Douglas, 161–162, 169, 175, 177, 181,
    199
Lewis, Sinclair, 200
Lewis v. Amorous, 87
Life, 221
Life-cycle costing, 132
Light bulbs. *See* Lamps/lights
Light-conditioning campaign, 150–151, *150*
Lighting celebrations, 43
Lighting schemes
  for industrial facilities, 107, 125–126, 132,
    143, *143*, *144*, 145, 232
  for schools, 107, 109, 146–147, *148*, 149
Lighting societies, 7
Light-mapping, *143*, *144*
Lightolier Co., *54*
"Light's On" (Auden), 237
Light switches. *See* Switches/buttons
Lissitzky, El, 185
Lists, 189, 192, 194
Literary Digest, The, 200
Literature, switches in, 53
Livermore, Mr., 123, 132
Locomotives
  braking distance of, 79–80
  lamps on, 68, *69*, 70
  night views from, 70, *71*, 80, 101
London, 157, 163, 194, 203, 209
Long Acre Square (New York), 159. *See also*
    Times Square (New York)
Long Island Rail Road, 163
Los Angeles (CA), 43, 201, *216*
Los Angeles County Highway Motor
    Patrol, 102
Luckiesh, Matthew, 126–127, 129, 137,
    149–150
Lucretius, 14

Madison Square Garden (New York), 159, 163
Madison Square (New York), 157, 163, *175*
Madness, 20
Madness and Civilization (Foucault), 20
Magdsick, H. H., 83
Magnetism, 14–15
Malevich, Kazimir, 9
Manchester (England), 118
Manchester (NH), 123
Man Ray, 185
Marx, Karl, 64, 118, 141
Matches, 3, 112
Material progress, electric lighting and, 7, 161, 209
Maxwell, James, 14–15
May, Mitchell, 84, 98
Mayakovsky, Vladimir, 11
Mayors, 42, 159
Mazda Lamps, *68*, *128*, *136*, *212*
McClellan, George, 159
McLuhan, Marshall, 20
McMurtry, Mr., 55
Measurement, of electric lighting, 149–151
Memory, 223
Memory value, 179–181, *180*
Mencken, H. L., 201, 203
Messmer, Otto, 175
Miller, Francis Trevelyan, 46
Mill fever, 120
Mills. *See* Textile mills
Milton Bradley Co., 237
Model T Ford, 72
Modernism
  in architecture, 2, 11–12, 51
  electric lighting and, 9, 11–12
Mollahan Mills (Newberry, SC), *117*
Monitorial System, 146
Montebello (CA), 43
Morale, 236
Morality, of better vision, 137, *139*, 140–143, 145
Moss, Frank, 149–150
Motion theory, 179
Motographs, 165, *166*, 167, 179
Mumford, Lewis, 126
Mundane, technological, 63–64
Münsterberg, Hugo, 179

Murrow, Edward R., 215, 217
Myopia, 147
Mystery stories, 235

Nader, Ralph, 151
National Electric Light Association (NELA), 48, *48*, 56, 59, 126, 149
National Paint, Varnish and Lacquer Association, 149
National Retailers Contractors Association, 149
National Retail Furniture Association, 149
NELA (National Electric Light Association), 48, *48*, 56, 59, 126, 149
Neon lighting, 161
New Haven (CT), *201*
New Republic, The, 194
News zipper (*New York Times*), 165, *166*, 167, 192
Newton, Isaac, 14
New York City
  headlights use in, 77
  lighting of
    Broadway (*see* Broadway [New York])
    and commercial speech (*see* Commercial speech)
    and Great White Ways (*see* Great White Ways)
    Times Square (*see* Times Square [New York])
  street lights in
    electric lighting of, 11, 159
    gas lighting of, 157
  World's Fairs in, 209, 217
New Yorker, 175
New York Times
  advertisements of, 163
  editorials of, 87
  moving to Times Square, 159
  news zipper, 165, *166*, 167, 192
  quotes from, 37, 221
  reporters of, 176
Niagara Falls, 197–200
Niblo's Garden, 157
Night blindness, 92, 232
Night driving. *See* Driving, at night
Nighttime labor. *See* Labor
Nostalgia, for darkness, 209, 211

Objects, affecting one another, 14
Offices, 151
Ohio Electric Light Association, 56
Oil lamps, 27, 28, 68, 112, 118
Oils, 2–3, 112
O. J. Gude Company, *160*, 161, *162*, 169
Olfactory senses, 222
Olympian Cotton Mills (Columbia, SC), *122*
Ophthalmological maladies, 92
Ordinariness, of switches, 63–64
Osgood lenses, 99, *99*
O'Shaughnessy, Edith Louise, 67
Owed to Your Eyes (film), 149

Pan (demigod), 233
Pan-American Exposition (Buffalo, NY), 198
Panic, 232–233
Panopticon prison, 152
Panoramic perception, 192
Paris, *4*, *5*, 51, 60–61, 157, 194, 203
Park & Tilford v. Realty Advertising, 187, 189–190
Parodies, on house electrical, 51–52, *52*
Parrish, Maxfield, *13*
"Path lighting," 35
Paul VI, Pope, 25
Pedestrians, 67, 72, *86*, 88, 92, *93*, 102
Perspective, rearrangement of, 193–194
Petticoats, Heatherbloom, 167, *168*
Petty, Margaret Maile, 7
Piccadilly Circus (London), 203
Pinewood, 112
Pius XI, Pope, 42
Place Pigalle (Paris), 203
Playtime (Tati), *52*
Pope, Franklin, 14
Popes, 25, 42
Posture, 147, *148*, 149
Presidents, 37, *38*, 39–40, 42, *42*, 43, 44, 45, *45*, 78, 94, 213, 217, 239
Prest-o-lite Co., 74
Price, C. W., 107
Privacy, right to, 33
Productive intensities, 135
Productivity
  electric lighting and, 125, 132–133, *134*, 135, 149
  electric motors and, 125

vision and, 125–127, *128*, 129
Prometheus (Parrish), *13*
Proust, Marcel, 61
Psychology and Industrial Efficiency (Münsterberg), 179
Publicity, 194
Pudong quarter (Shanghai), 204

Radio broadcasts, 215, 217, 232–233, 240
Railroads. *See* Locomotives
Rand, Gertrude, 232
REA (Rural Electrification Administration), 43, 45
Reaction time, 80
Reflectivity, *93*, 207, *226*, 229
Reflectors, 68
Regulations
  on automobiles, 81
  on blackouts, 229, 231
  on headlamps
    municipal, 83, 92–93
    standardization of, 97, 100
    variations in, 92–93
Rehearsals, of blackouts, 239
Reilly, Frank C., 163, 165, 179
Repetition, 179
Replogle, J. B., 94
Révész, Géza, 222
Rice Electrical Company, 169, *173*
Riley, James, 32
Rio de Janeiro, 25
Roadbuilding, 94
Rockland (ME), 43
Rockwell, Norman, *33*, *234*, 235
Rooms, electric lighting of, 59–60
Roosevelt, Franklin D., 42, 43, 44, 45, 213, 217, 239
Roscoe (Sun Valley, CA), 43
Rothacker Film Manufacturing Co., 59
Rothschild, Philippe de, 203
Routines, 236–237
Rowlandson, George, 51
Rumely Products Co., *36*
Rural Electrification Act, 209
Rural Electrification Administration (REA), 43, 45, *210*, 211
Rushlights, 112

Safety, 81, 140
Salt Lake City (UT), *219*
San Francisco, 46
Saturday Evening Post, 233, *234*, 235
Scheerbart, Paul, 11
Schivelbusch, Wolfgang, 192, 200
Schools
  electric lighting of, 107, 109, 146–147, *148*, 149
  as factories, 107, 109, 146
Schrecklichkeit, 233
Schudson, Michael, 194–195
Scientific American, 70
Scotopic space, 225, 227, *227*, 230, 233, 236
Scott, Charles Felton, 132
Scripture, Dr., 146
Second Congress on Large Dams, 43
Second National Swiss Automobile Exposition (Geneva), *90*, 91
See Better–Work Better Bulletin, 151
Seeing. *See* Vision/seeing
Self-control, 34
Self-propulsion, 105
Senses, *220*, 221–223, 225
Servants, 55
Seymour, Robert, 51
Shanghai, 203
Shape of Things to Come, The (Wells), 215
Sheffield Scientific School, 133, 135
Sheridan, Ann, 91
Shirer, William L., 213, 221
Shops, 146
Signs of the Times, 185
Simons, Eric Norman, 191
Skin, 223, *225*
Slauson, Harold Whiting, 75, 77
SMILAX, 155
Smith, Adam, 113
Somaesthetics, 225
Somatic attention, 225
Son et lumière spectacles, 176
Space
  of automobile headlamps, *81*, *82*, *85*, 102, 103, *104*
  light switches and, 34–35, 59–60
  mechanical motion and, 105
  perception of
    braking distance and, 79

memory and, 223
senses and, 221–222
scotopic, 225, 227, *227*, 230, 233, 236
Speed
  efficiency and, 126–127
  night driving and, 83, 87–88, 103
  of vision, 126–127, *127*, *128*, 129, 149
Spine injuries, 147, *148*, 149
Stamps, 6
Standardization, 136–137, *136*
Star, Jacob, 161
Star Wars (films), 203
Steinmetz, Charles, 18
Stereopticon, 176
Stevenson, Robert Louis, 17, 161
St. Gaudens, Augustus, 159
Stimulating effects, of electric lighting, 142
Strauss, Benjamin, 161
Street Light (Balla), *10*, 11
Streets
  as cinema, 176, 177–178
  electric lighting of
    in general, 5, 200–201
    in New York City, 11, 159
    in Paris, *4*, 5
    in San Francisco, 46
  gas lighting of
    in New York City, 157
Strunsky, Simeon, 199
Sublime, 196–197, 198
  technological, 63
Supervision, 145
Swan, Joseph, 5
Switches/buttons
  action at a distance of, 49
  agency and, 35, 60, 61
  appearance of, 28
  device paradigm and, 49
  in domestic settings, 49, *50*, 51–53
  failure to find, 63–64
  familiarity of, 63–64
  in films, 51–52, *52*
  flipping/pressing of
    by Cleveland, Grover, 37, *38*, 39–40
    by Kennedy, John F., 45
    magic of, 46, *47*, *57*
    on portable cameras, 40
    by Roosevelt, Franklin, 43, *44*, 45

INDEX   289

Switches/buttons (cont.)
  by Truman, Harry, 42
  by women, 53, *54*, 55–56, *58*
  housework and, 51
  ignorance and, 60
  instantaneity of, 49, 60, 61, *62*, 63
  light
    appearance of, 28
    conception of, 27
    control and, 33, 36
    designs of, 28, *29*, 31, *31*
    device paradigm and, 49
    in general, 15, 26, 36
    operation of, 32–33
    and space, 34–35, 59–60
    volitional power of, 34, 37, 39–40
  in literature, 53
  meanings of, 27
  ordinariness of, 63–64
  as symbol of progress, 45
  volitional power of, 51, 60

Tailors, 109
Tally-Ho Bicycle Lanterns, *69*
Tati, Jacques, *52*
Taut, Bruno, 11
Taxonomies, 189
Taylor, Frederick Winslow, 129, 141
Taylor, William R., 155
Technology
  device paradigm and, 46–47
  female incompetence with, 56
Telegraphy, 15
Temperatures, at textile mills, 120
Tesla, Nikola, 132
Texas Company, 181
Textile industry, 113. *See also* Textile mills
Textile mills
  Amoskeag Manufacturing Company, 123, 132
  Belt-and-shaft systems in, 113
  Boott Cotton Mills (Lowell, MA), *115*
  Cromford Mill (Derbyshire), 118, *119*
  design of, *115*
  evening/nights shifts at, 118, 120
  lighting of
    by candles/oil lamps, 118
    by daylight, 113–114, *115*, 117

  by electric light, 120–121, *122*, 123–124
  by gas lamps, 118, *119*, 120, 121
  in Manchester, 118
  in the Midlands, 118
  Mollahan Mills (Newberry, SC), *117*
  Olympian Cotton Mills (Columbia, SC), *122*
  temperatures at, 120
They Drive by Night (film), 91
Third World Power Conference, 43
"This Space for Rent" (Benjamin), 196
Thompson, E. P., 145
Thurber, James, 32
Tide, 167
Times Square (New York). *See also* Broadway (New York)
  during blackout, *224*
  lighted commercial speech in
    and Adams, Mildred, 194–195, 209, 211
    attention/memory value and, 180–181
    and Bennett, Arnold, 186
    and Berman, Marshall, 155, 193, 199
    and Brooke, Rupert, 186–187
    and Chesterton, G. K., 187, 191
    as cinema, 177
    compulsory aspect of, 185
    and Crane, Stephen, 159
    and Evans, Walker, 184
    in general, 155, *156*, *158*, 176, *195*
    and Gude, Oscar J., 159, *160*, 161, *162*, 163, 169, 177, 198–199
    and Hughes, Rupert, 159, 186
    and Leigh, Douglas, 161–162, 174, 199, *199*
    meanings of, 194–196
    Niagara Falls and, 197–200, *199*
    perception of, 192–194
    picture animation of, 167, *168*, 169, *170–171*
    and Rothschild, Philippe de, 203
    and Schivelbusch, Wolfgang, 200
    and Strunsky, Simeon, 199
    sublime experience and, 196–197, 198
    text animation of, 163
    urban reading of, 181, *182*, 183, *183*, 185–187, *188*, 190–192
    and Wells, H. G., 198
Tokyo, *204*
Touch, sense of, 222

Trains. *See* Locomotives
Trimble Whiskey sign, 177
Tripper, Harry, 180–181
Tropicana products, 169
Truman, Harry, *42*
Tunnel vision, 91–92
Turner, Frederick Jackson, 181

Union Square (New York), 157, 159
United States. *See also under specific cities*
 blackouts in, *71*, 207, *208*, 215, *216*, 217, *218–219*, *224*, 231–232, *234*, 236, 238–240
 Great White Ways in, 200–201, 203
 panic, spread of, 232–233
 street lighting in, 5, 11, 46, 157, 159
University of Pennsylvania, 135
Urban reading, 181, *182*, 183, *183*, 185–187, *188*, 190–192, *202*
USDA Housekeepers' Chat Radio Service, 149

Value
 of attention/memory, 178–181, *180*
 of electric lighting, 132–133, *134*, 135
Valves, 27
Van Dusen, Mr., 176
Van Dyke, Henry, 68
Villa Feria Electra, 51
Vision/seeing. *See also* Eyes
 apprehension and, 126–127, *127*, *128*, 129
 as commodity, 107
 fatigue of, 141–142
 industrial rationale about, 151–152
 morality of better, 137, *139*, 140–143, 145
 night driving and, 73, *73*, *74*
 nighttime
  eyes and, *220*, 221–222
  memory and, 223
  nearness and, 230
  other senses and, 222
 posture and, 147, *148*, 149
 productivity and, 125–127, *128*, 129
 speed of, 126–127, *127*, *128*, 129, 149
 supremacy of, 211
 of Times Square, 192–194
 tunnel, 91–92

Volitional power, of switches, 34, 37, 39–40, 51, 60
Volitional spaces, 34–35

Wachtel, Curt, 222
Walker, John, 3
"War of the Worlds" (Welles; radio broadcast), 232–233
Warren, Dean M., 112
Warren, Samuel, 33
Washington Post, 220, 233, 236
Welles, Orson, 232–233
Wells, H. G., 196, 198, 215
Welsbach, Carl Auer von, 121
Welsbach gas mantle, 7
Westinghouse, George, 5
Westinghouse Individual Motor Drive, *124*
Westinghouse Lamps Co., *134*
White, Ethel Lina, 235
White, Stanford, 159
White Rock Table Water, 167, 169
White way lighting, 200–201, 203
Williams, Edward T., 196
Williams, Raymond, 145
Wilson, Woodrow, 42, 78, 94
Winchell, Walter, 162
Women
 driving by, 91–92, *104*
 electricity and
  explaining to, 56, 59
  incompetence with, 56
 glare and, 89, 91
 joyriding and, 89
 nighttime labor by, 109, *110*, 112
 pushing buttons by, 53, *54*, 55–56, *58*
Women's Public Information Committee, 56
Woodbury, Charles Jeptha Hill, 121, 123, 126, 132–133
Woolf, Virginia, 61
Wordsworth, William, 194
Workers. *See* Laborers
World's Columbian Exposition (Chicago), 37, *38*, 39–40, 163
World's Fair (Chicago), 37, *38*, 39–40, 163
World's Fair (New York), 209, 217
World War I, 135, 209, 213, 232

World War II, 207, 215, 217
Wright, Frank Lloyd, 59
Wrigley's Spearmint Chewing Gum, 169, *170–171*
Wundt, William, 179

Yablochkov, Pavel, *4, 5*
Yale University, 133, 146

Zola, Émile, 18–19, 68